ISBN 978-1-330-63706-7
PIBN 10085698

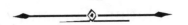

English
Français
Deutsche
Italiano
Español
Português

www.forgottenbooks.com

Mythology Photography **Fiction**
Fishing Christianity **Art** Cooking
Essays Buddhism Freemasonry
Medicine **Biology** Music **Ancient
Egypt** Evolution Carpentry Physics
Dance Geology **Mathematics** Fitness
Shakespeare **Folklore** Yoga Marketing
Confidence Immortality Biographies
Poetry **Psychology** Witchcraft
Electronics Chemistry History **Law**
Accounting **Philosophy** Anthropology
Alchemy Drama Quantum Mechanics
Atheism Sexual Health **Ancient History**
Entrepreneurship Languages Sport
Paleontology Needlework Islam
Metaphysics Investment Archaeology
Parenting Statistics Criminology
Motivational

THE

LONDON DISSECTOR;

OR,

SYSTEM OF DISSECTION,

PRACTISED IN THE

HOSPITALS AND LECTURE ROOMS,

OF THE METROPOLIS,

EXPLAINED BY THE CLEAREST RULES,

FOR THE USE OF STUDENTS;

COMPRISING

A DESCRIPTION OF THE MUSCLES, VESSELS, NERVES, AND VISCERA, OF THE HUMAN BODY, AS THEY APPEAR ON DISSECTION; WITH DIRECTIONS FOR THEIR DEMONSTRATION.

BY JAMES SCRATCHLEY,

SURGEON TO THE ROYAL REGIMENT OF ARTILLERY,
AND TO THE CORPS OF ROYAL ENGINEERS.

EIGHTH EDITION, REVISED.

LONDON:

THOMAS AND GEORGE UNDERWOOD,

FLEET STREET.

1829.

494

LONDON:

PRINTED BY R. GILBERT,
ST. JOHN'S SQUARE.

PREFACE.

THE Writer of the following pages, having frequently witnessed the difficulties which attend the pursuit of Practical Anatomy, conceived that some assistance might be derived from a Compendium, describing the various parts of the Human Body, as they come into view under the knife of the Dissector. With this intention the present performance is offered to the Public.

The muscles are demonstrated in the order of their situation, this being the only method that can be pursued in actual Dissection.

The relative situation of the several parts is minutely attended to, and, at the same time, repetitions are avoided, as far as the nature of the arrangement adopted would allow. Should this work in any degree facilitate the progress of the Anatomical Student, the object of the Author will be attained.

London, 1804.

ADVERTISEMENT

TO THE

SEVENTH EDITION.

———

THE uninterrupted sale, during twenty years, of large editions of the London Dissector, and the still continued demand for the book, have sufficiently shewn, that it has, in some degree, fulfilled the intention of the Writer, in supplying to the Student a convenient Manual of Practical Anatomy, which, at the time of its first publication, was a work much wanted in the Dissecting-room. In the present edition, the Author has revised generally his little Volume, verifying anew the more important descriptions by numerous recent dissections, and introducing many corrections and additions.

Royal Ordnance Hospital, Woolwich,
August, 1826.

CONTENTS.

A 3

CHAPTER II.

CHAPTER III.

CHAPTER IV.

DISSECTION OF THE THIGH.

CHAPTER V.

DISSECTION OF THE LEG AND FOOT.

SECT. I. OF THE FORE PART OF THE LEG AND FOOT.

CHAPTER VI.

Dissection of the Head.

CHAPTER VII.

DISSECTION OF THE ANTERIOR PART OF THE NECK.

CHAPTER VIII.

DISSECTION OF THE THORAX.

CHAPTER IX.

DISSECTION OF THE FACE.

CHAPTER X.

DISSECTION OF THE NOSE, MOUTH, AND THROAT.

CHAPTER XI.

DISSECTION OF THE EYE AND ITS APPENDAGES.

CHAPTER XII.

CHAPTER XIII.

CHAPTER XIV.

CHAPTER XV.

CHAPTER XVI.

PRACTICAL ANATOMY.

GENERAL RULES FOR DISSECTION:

DEXTERITY in the manual operation of dissection can only be acquired by practice ; the observance, however, of certain general rules, will facilitate the labour of the student.

1. The position of the hand in dissecting should be the same, as in writing or drawing ; and the knife, held, like the pen or pencil, by the thumb and the two first fingers, should be moved by means of them only ; while the hand rests firmly on the two other fingers bent inwards as in writing, and on the wrist. The instrument can be guided with much more steadiness and precision in this way, than when it is moved by means of the wrist, elbow, or shoulder, in the manner which young dissectors often fall into.

2. No more of the integuments should at any time be removed, than is necessary for the present dissection, as exposure to the air renders the parts dry and indistinct.

B

3. In dissecting muscular parts, the muscles should be extended; the cellular membrane which connects them to the integuments, should be placed on the stretch; and entirely removed with the skin; the knife should be kept close to the muscles, and carried steadily in the direction of their fibres, separating a fasciculus at each stroke: thus the exposed surface will appear clean, and the course of the fibres distinct.

4. When small vessels are to be demonstrated, another method is to be followed; the skin only must be removed, and the cellular membrane cautiously and slowly dissected from the vessels.

5. During dissection, every little operation should be practised, which can give the dexterity of hand so essential to the surgeon; such are, the use of the catheter and probang, the introduction of a probe through the nose into the Eustachian tube, or nasal duct, and the cutting down to the various arteries, which may become the object of surgical operations: as the external iliac, femoral, anterior and posterior tibial, brachial, radial, and ulnar, &c.

The grand object of the surgical student is to acquire a knowledge of the relative situation of parts. This should be kept in view in all his anatomical labours. Hence, when he is dissecting the muscles, he should carefully expose the chief blood vessels and nerves; and attentively consider their position with regard to each other, and to the surrounding parts. This species of knowledge will afford him the most essential assistance in his future operations on the living subject; in which indeed it is so necessary, that we are perfectly astonished to see persons rash enough to use the knife without possessing this information: but we view the hesitation, confusion and blunders, by which such operators betray their ignorance to the bystander, as the natural result, and the well-merited but too light punishment, of such criminal temerity. The smaller arteries and veins, and the minute nervous ramifications, will be more advantageously studied in subjects devoted to those purposes, and prepared by means of injection, immersion in spirits of wine, &c.

The reader will observé, that, in general, the muscles of one side of the body only are described, because all the muscles of the body have corresponding ones on the opposite side, with a few exceptions which are pointed out.

CHAPTER I.

DISSECTION OF THE ABDOMEN.

———

IN dissecting a subject, it is usual to begin with the MUSCLES of the ABDOMEN.

SECTION I.

MUSCLES OF THE ABDOMEN, AND THE PARTS CONNECTED WITH THEM IN DISSECTION.

THESE muscles are ten in number, five on each side.

Place a block under the loins, to put the muscles on the stretch ; make an incision through the integuments from the sternum to the os pubis, cross it by another passing obliquely upwards from the umbilicus over the cartilages of the ribs ; and dissect off the flaps :—this will lay bare,

1. The OBLIQUUS DESCENDENS EXTERNUS.—*Origin :* By eight triangular fleshy slips from the lower edges and external surfaces of the eight inferior ribs, at a little dis-tance from their cartilages ; the five superior slips meet on the ribs an equal number of the digitations of the serratus magnus, and the three inferior are connected with the attachments which the latissimus dorsi has to the ribs. Often there are only seven portions. To gain a complete view of this muscle, the neighbouring por-tions of the pectoralis major, serratus magnus, and latis-simus dorsi should be dissected with it.

The muscular fibres proceed obliquely downwards and forwards, and, about the middl of the side of the belly, terminate abruptly in a thin broad tendon, which is continued in the same direction over all the fore part of the belly. Here it covers the anterior surface of the rectus abdominis; it is very thin at the upper part, where the rectus lies on the cartilages of the ribs, and is often removed by the beginner, unless he is very cautious.

Insertion: Tendinous and fleshy, into two anterior thirds of the outer edge of the crista of the os ilium, from the anterior superior spine of which it extends obliquely downwards and forwards to the os pubis, forming Poupart's ligament; into the ensiform cartilage; and into the whole length of the linea alba, where it meets and is united with the tendon of the opposite side.

Situation: It is quite superficial, and covers the whole of the anterior part of the abdomen, forming the first or external layer of the fleshy walls. The muscular part is closely covered by a thin expansion of cellular substance, which, as it descends over the tendon, becomes gradually thicker, and, at the lower part of the abdomen, has the appearance of a distinct fascia, and is named the *Fascia Superficialis.* The posterior free border of this muscle, descending nearly vertically from the last rib to the crista ilii, ranges with the anterior edge of the latissimus dorsi, and is sometimes overlapped by it. A slender artery is constantly found ascending over the tendinous expansion of the obliquus externus; this is the *External* or *Superficial Epigastric* artery, which comes off from the femoral artery in the groin, passes over Poupart's ligament, and may be traced as far as the umbilical region.

Use: To draw down the ribs in expiration, to bend the trunk forwards when both muscles act, or obliquely to one side when one of them acts singly; to raise the pelvis obliquely when the ribs are fixed; and to compress the abdominal viscera.

In the course of the dissection of this single muscle, the following points must be attended to.

The FASCIA SUPERFICIALIS of the abdomen. This is generally removed with the integuments, otherwise the dissection of the muscle is not clean and neat ; but it may always be demonstrated. At its upper part, where it covers the muscular fibres of the obliquus externus, it consists chiefly of condensed cellular tissue, being continuous with a similar texture investing the pectoralis major; lower down it becomes thicker, and, over the tendinous part of the obliquus externus, it is quite distinct as an aponeurotic expansion, and there is generally more or less of adipose substance betwixt this fascia and the subjacent tendon. It passes on to the thigh over Poupart's ligament, and is finally lost in the fascia lata of the thigh. It also covers the abdominal ring, and passes over the spermatic chord into the scrotum, and here it may be easily separated by the handle of the scalpel, or by the blow-pipe ; it also passes on to the dorsum of the penis. In the female, it is continued into the labia.

The LINEA ALBA, a white, dense, tendinous line running down the middle of the abdomen, from the cartilago ensiformis to the os pubis : formed by the tendinous fibres of the two obliqui and the transversalis muscles, interlaced with those of the same muscles on the opposite side ; it is half an inch broad at the navel; and decreases gradually both above and below that part; but particularly in the latter situation, where it is reduced at last to a mere line.

LINEA SEMILUNARIS, a semicircular white line, running from the os pubis obliquely upwards over the side of the abdomen, at the distance of about four inches from the linea alba; formed by the tendons of the two line of union for several important fasciæ, and, at the present stage of the dissection, its connexion with the fascia lata of the thigh may be advantageously studied. I would therefore advise the student, before he proceeds to the dissection of the other abdominal muscles, to dissect the parts in the groin, and to examine the situation of the great inguinal vessels.

In taking off the skin from the groin, you find a confused and irregular Aponeurosis, coming off from the abdomen, and going down upon the thigh: this is a continuation of the fascia superficialis, already described, as covering the expanded tendon of the obliquus externus; and it lies exterior to the fascia lata of the thigh. This *Superficial Fascia* may be observed descending from the abdomen over Poupart's ligament, with which it is connected, and more strongly so with the iliac portion of the ligament: at the hollow of the groin, it closely invests, and is interlaced with the absorbent glands, and adipose and cellular tissue; and it may be traced a considerable distance down the thigh, especially in emaciated subjects, until it is lost in the fascia lata and integuments.

Under this superficial fascia lies the *Fascia lata* of the thigh: if we commence exposing the fascia lata on the outer part of the limb, in a line with the spine of the ilium, we find it thick and strong, but as we advance to the groin, and inner side of the thigh, it becomes much thinner. The dissection is to be prosecuted by removing cautiously the superficial fascia, together with the mass of fat, glands, and cellular substance from the hollow of the groin; and as we proceed in clearing away this confused mass, the fascia lata will be seen to assume the form of a crescent, the concavity of which is turned to the other thigh: this is the *Falciform Process*, or *Semi-*

lunar edge of the fascia lata ; and here the *Vena Saphena,* after running up the thigh, is seen to dip down, and terminate in the *Femoral* or *Inguinal Vein,* about an inch and a half below Poupart's ligament.

The FASCIA LATA on the upper and fore part of the thigh, with its attachments, may now be fully exposed, and, for the convenience of description, it may be distinguished into two portions or divisions. The outer, more dense, or *iliac* portion is seen covering the muscles on the outer part of the thigh, adhering to the spine of the ilium, and to Poupart's ligament as far as within an inch of the pubis ; at this point the fascia ceases to be attached to the ligament, and, by receding downwards, forms the Falciform process, the sharp edge * of which will be found to be continued across the thigh, behind the vena saphena, into the inner and thinner portion of the fascia lata, which is seen ascending over and investing the gracilis and pectineus muscles, extending up to their pubal attachments, and firmly attaching itself to the os pubis ; this inner, or *pubic* portion of the fascia lata is more deeply seated, and is connected with the sheath of the great vessels, and passes behind these vessels to become continuous, over the brim of the pelvis, with the iliac fascia : an *oval aperture,* or *interval* is thus formed between the two portions of the fascia lata, and, at this point, the *Vena Saphena* is seen entering to join the Inguinal Vein, and the great Inguinal Vein itself is here not covered by the fascia lata of the thigh, but only by the superficial fascia, glands, and cellular tissue, and by its own proper sheath.

POUPART'S LIGAMENT or the *Crural Arch,* is now

* This edge or margin of the falciform process is found, on closer examination, to be inflected inwards, and to become connected with the sheath of the vessels ; in subjects loaded with fat the fasciæ are less distinct.

more fully exposed, and it will be observed that it is formed by a fold or doubling back of the tendon of the external oblique. If we now lift up Poupart's ligament, we shall perceive that its lower edge, near the pubis, is connected with that portion of the fascia lata, which invests the pectineus muscle, and with the sheath of the great vessels, by a thinner membranous portion, which, being perforated by several small foramina, has received the name of the *Cribriform Lamella*, or *Cribriform Fascia.*

By a little further dissection, we come down upon the *Great Vessels;* we find them closely invested by a firm sheath, which is much connected with the several fasciæ. The remaining cellular tissue and lymphatic glands, together with that portion of the fascia lata, which arises from Poupart's ligament anterior to the vessels, should be cautiously removed with the scissars *; the triangular space which is formed at the hollow of the groin, between the sartorius and adductor muscles, is now fully displayed : it is bounded above by the line of the crural arch. Observe the order in which the parts are situated under the crural arch :—that the great *external Iliac Vein* lies next to the pubis ; that the *external Iliac Artery* is close on the outside of the vein, invested in the same sheath, but separated by a distinct tendinous septum ;—that the *anterior Crural Nerve* is half an inch exterior to the artery, and lies on the iliacus internus ; and that the outer half of the space left under the crural arch is filled by the psoas magnus and iliacus internus muscles. Observe how the *Crural Arch* is stretched over the parts just enumerated ;—how, after being inserted by a roundish tendon into the angle or spine of

* In taking off the falciform process of the fascia lata, we remark that it passes in a twisted form under Poupart's ligament ; it will be found to be continued into Gimbernat's ligament.

the os pubis, it extends from that point backwards along the crista of the bone, with a thin, but firm, sharp, and crescent-shaped edge, constituting the ordinary seat of stricture in the femoral hernia, and sometimes called *Gimbernat's Ligament* * ; observe that the direction of this ligament of Gimbernat is nearly horizontal in the erect posture, and that its extent is about an inch or more, and that it forms the inner boundary of the space or passage by which the great vessels leave the abdomen. Observe also that, at the inner side of the passage for the great vessels, near the pubis, there is a space between the iliac vein and Gimbernat's ligament, only occupied by cellular tissue and some lymphatics, and through which femoral hernia protrudes : this aperture has been named the *Femoral* or *Crural Ring* †, and it leads from the abdomen into the thigh upon the fascia lata covering the pectineus. Observe how the great iliac artery is placed in relation to the angle of the pubis and spine of the ilium :—how it passes under the middle of the arch, or rather nearer to the os pubis than to the ilium, and, before it has emerged fairly into the thigh, sends off two considerable branches.

(1.) A. CIRCUMFLEXA ILII is sent off from the outer side, and passes upwards and outwards, running along the inside of Poupart's ligament, until it arrives at the anterior superior spinous process of the os ilium; near its origin, it pierces the iliac fascia, which fascia, in its line of attachment to Poupart's ligament, forms a sheath or slender channel for this artery. At the spine of the ilium it usually subdivides, one branch ascending between the transversalis and internal oblique muscles,

---

* Or, by some, the third insertion of the external oblique.

† Gimbernat's ligament and the Crural Ring are more readily seen within the abdomen. See Section II. of the present chapter.

while the other, larger, branch follows the curve of the crista ilii, giving twigs to the iliacus internus, and inosculating with the ilio-lumbar artery. That distribution, however, cannot be seen in this stage of the dissection.

(2.) The EPIGASTRIC ARTERY comes off from the inside of the iliac; it often descends a little in a tortuous manner, then crosses the external iliac vein, passes obliquely upwards and inwards, under Poupart's ligament, to which it is but loosely connected, and crosses behind the spermatic chord, to the outer edge of the rectus muscle. At first it is situated between the fascia transversalis and the peritoneum; then between the posterior surface of the rectus and the peritoneum; but higher up between the muscular fibres and the sheath. In a future stage of this dissection, the precise situation of the epigastric artery will be found to be close to the lower or pubic margin of the internal ring, and at some distance above the upper extremity of the external ring. At this point, it gives off an artery, which passes on the spermatic chord, or round ligament, and one or two smaller vessels, which cross horizontally behind Gimbernat's ligament and posterior surface of the pubis : then reaching the edge of the rectus, and continuing to ascend, the trunk of the epigastric terminates in ramifications, which supply the rectus and neighbouring muscles, and which inosculate with the mammaria interna, intercostal and other arteries. It is accompanied by two veins, which come from the iliac vein.

The Epigastric sometimes arises by a common trunk with the obturator artery. The point at which it springs from the iliac is also variable; it has been seen to arise from the femoral artery.

The dissection of the other abdominal muscles may now be continued.

Dissect off the serrated origin of the external oblique from the ribs, and from the spine of the os ilium, and detach it from the obliquus internus, which lies below it, and which is connected to it by loose cellular substance, and by small vessels. Continue to separate the two muscles, till you find their tendons firmly attached, *i. e.* a little way beyond the linea semilunaris. Also carefully divide the expanded tendon along the line of Poupart's ligament, from the ilium downwards to within half an inch of the external ring. This will expose a muscle, the fibres of which are seen passing upwards, crossing in an opposite direction.

2. OBLIQUUS ASCENDENS INTERNUS.—*Arises* by short tendinous fibres, which soon become fleshy, from the whole length of the spine of the os ilium, and from the fascia lumborum *; also fleshy from the outer half, or rather more, of Poupart's ligament internally.

The fibres run in a radiated direction; those which originate from the lumbar fascia and back part of the os ilium, run obliquely upwards; those from the fore part of the ilium pass more transversely across the belly; and from Poupart's ligament the fibres descend. The fleshy belly is continued rather more forward than that of the external oblique, before it terminates in a flat tendon.

Inserted into the cartilages of the six or seven lower ribs,—fleshy into the three inferior, and, by a tendinous expansion, which is extremely thin, resembling cellular membrane, into the four superior, and also into the ensi-

* Some describe it as arising from the sacrum and three inferior lumbar vertebræ; but this is not accurate. It arises from a tendinous fascia common to it and the next muscle, and to certain muscles of the back, as the serratus posticus inferior and latissimus dorsi; this may with propriety be named *Fascia Lumborum.*

form cartilage. The sheet of tendon in which the fleshy belly ends, is continued, single and undivided, into the linea semilunaris, where adhering pretty firmly to the tendons of the obliquus externus and transversalis, it divides into two layers. The anterior and more considerable layer joins the tendon of the external oblique, and runs over the rectus to be inserted into the whole length of the linea alba : the posterior and thinner layer, adhering to the anterior surface of the transversalis, passes into the linea alba behind the rectus, as low as midway between the umbilicus and os pubis; but below this place, the whole tendon of the internal oblique passes, along with that of the external oblique, before the rectus, and is inserted into the lower part of the linea alba. The inferior edge of the muscle extends in a nearly straight direction over the spermatic chord, to be fixed by a tendinous attachment, common to it and the transversalis, into the angle of the pubis.

Situation : This is the second of the muscular layers of the belly. It is covered by the obliquus descendens externus and latissimus dorsi : the tendinous attachment to the os pubis is immediately behind the external ring, and may be felt by passing the finger into the ring, above the spermatic chord, and it serves to close the abdominal cavity in that direction.

Use : To assist the obliquus externus : but it bends the trunk in the reverse direction, so that the muscle on each side co-operates with the obliquus externus of the opposite side.

About the middle of Poupart's ligament, a delicate fasciculus of fibres is sent off from this muscle over the spermatic chord, where it passes under its edge in its way to the ring. This is named the

CREMASTER, and is continued down on the chord, till

it is insensibly lost on the tunica vaginalis testis ; it will be seen in the dissection of the scrotum ; Its *use* is to suspend, draw up, and compress the testicle.

We must now dissect the attachments of the internal oblique from the cartilages of the ribs, from the fascia lumborum, and from the spine of the os ilium, and, by continuing our dissection from behind forwards, separate it from the transversalis abdominis which lies under it. This separation may be continued as far as where the tendons of the two muscles are inseparable, *i. e.* rather more forward than the linea semilunaris. As this muscle lies very close upon the transversalis, caution is required to avoid detaching both muscles together. Let the student begin his separation at the crista of the ilium, where a branch of the circumflexa ilii artery and vein, and some cellular tissue, will shew him when he has arrived at the surface of the transversalis ; at its lower part, the obliquus internus is so closely connected with the transversalis, that a distinct separation is not easily effected.

3. Transversalis Abdominis—*Arises*, above, from the internal surface of the six or seven lower ribs, by digitations, which intermix with those of the diaphragm ; behind, from the fascia lumborum, and by a broad tendon from the transverse processes of the lumbar vertebræ * ; below, by fleshy fibres from two anterior thirds

* The posterior tendinous origin of this muscle is by three distinct *lamellæ*, or layers : (1) It springs from the fascia lumborum, in conjunction with the internal oblique : (2 and 3) the proper tendon of the transversalis splits into two laminæ, of which the *posterior* and strongest passes between the mass of sacro lumbalis, and the quadratus lumborum, to be fixed to the lower border of the last rib, and to the apices of the transverse processes of the lumbar vertebræ ; while the *anterior*, thinner, lamina passes in front of the quadratus, and is attached to the roots of the same processes.

of the inner labium of the crista ilii, and from the outer half of Poupart's ligament.

The fleshy fibres proceed transversely, and end in a flat sheet of tendon, which, after being connected to the other tendons at the linea semilunaris, passes with the posterior layer of the internal oblique behind the rectus, and is *inserted* into the ensiform cartilage, and into the whole length of the linea alba, excepting its lowermost part; for, at the middle distance between the umbilicus and os pubis, a slit or fissure is formed in this tendon, through which the rectus abdominis passes; and the remainder of the tendon passes before the rectus, to be inserted into the lower part of the linea alba. Its inferior edge is connected with that of the preceding muscle in its insertion into the pubis; the fibres also of the two muscles, where they arise from Poupart's ligament, are much intermixed: they arise from the ligament, rather more than half its length.

Situation: This forms the third and innermost layer of the broad abdominal muscles. It is covered by the obliquus internus, and is lined by the fascia transversalis and peritoneum.

Use: To support and compress the viscera of the abdomen.

When the transversalis is detached from its origins, and turned back towards the linea semilunaris, the peritoneum is laid bare, except in the neighbourhood of the crural arch, where it is covered by a thin fascia arising from Poupart's ligament, and continued upwards between the peritoneum and transversalis muscle, until it is gradually lost. This is named by Sir Astley Cooper, its discoverer, the *fascia transversalis*, and it prevents the bowels from being protruded under the inferior margins of the obliquus internus and transversalis muscles.

It is perforated midway between the spine of the ilium and the symphysis pubis, and about half an inch above Poupart's ligament, by an opening for the passage of the spermatic chord, which then goes obliquely downwards, inwards, and forwards, to the ring of the external oblique.

To discover the fascia transversalis, separate the fibres of the transversalis muscle from Poupart's ligament, and take them off cautiously from the subjacent fascia ; or open the abdomen, and strip off the peritoneum from its inner surface. This fascia will be found to arise from the posterior margin of Poupart's ligament in the whole of its length; but it varies much in density. Sometimes it is seen as a distinct fibrous expansion, springing from the spine of the ilium and the line of Poupart's ligament, adhering intimately to the lower transverse fibres of the transversalis muscle, and fixed to the outer edge of the rectus tendon : in others, it forms a tissue nearly cellular, but, by pulling on the spermatic chord, the margins of the aperture, by which it leaves the abdomen, can always be made out ; usually, the inner margin or pillar is falciform and more distinct, and the edges of the opening appear prolonged on the chord, in its passage through the abdominal walls. The fascia transversalis will be found continuous, along the line of Poupart's ligament, with the iliac fascia, and also connected with the sheath of the femoral vessels anteriorly.

The *Abdominal Ring* is, therefore, strictly a canal, having an upper or internal opening, the *internal ring*, formed in the fascia transversalis ; and a lower or *external* one, in the tendon of the external oblique ; and the whole passage is termed the *Inguinal Canal*. This canal is from one and a half to two inches in length. If the abdomen has been opened, the entire tract or passage of the *spermatic chord* may now be traced : within the abdomen, the chord is seen behind the peritoneum, formed of the vas deferens ascending from the pelvis, and of the spermatic vessels and nerves descending from the spine : if we gently pull the chord, the point at which it passes through the fascia transversalis is made evident by a pit or depression : this marks the situation of the internal ring, and the epigastric artery is seen ascending from behind Poupart's ligament, and passing upwards close to the inner or pubic edge of the internal ring, crossing the line of the spermatic chord, but situated behind it. Having left the abdominal cavity by this aperture, the chord runs under the edge of the obliquus internus and transversalis muscles, which pass over it like an arch, and, the fibres of the two muscles being much intermixed, the transversalis seems occasionally to contribute to the formation of the cremaster * : the chord is here situated about half an inch above the line of the crural arch ; behind it is the fascia transversalis, while anteriorly it is covered by the tendon of the obliquus externus : it descends obliquely, approaching nearer to the crural arch in its descent, and being lodged in a kind of groove or half canal, formed by the doubling back of the tendon : it then emerges through the external ring : it lies close to the external or lower column of the ring, and, becoming superficial, turns down over the pubis into the scrotum. In this oblique passage through the inguinal canal, the chord is invested, successively, by the funnel-like prolongation of the transversalis fascia, by the cremaster muscle, and, finally, on

---

* In some subjects, the fibres of the two muscles, and particularly of the internal oblique, continue to arise from the line of Poupart's ligament further downwards, and the spermatic chord then passes between the muscular fibres, which open to allow its passage.

passing through the external ring, by the fascia propria, superficial fascia and common integuments.

In the female, the Inguinal Canal is much less developed ; the opening in the transversalis fascia is small, and it is filled up by the *ligamentum teres,* a fibrous fasciculus, which comes forward from the uterus, in the anterior fold of the broad ligament, and is ultimately lost in the fatty tissue of the mons veneris and labium. There is nothing analogous to the cremaster.

The sheath of the rectus is now to be attended to; it is formed by the tendons of the three other muscles, viz. the two obliqui, and the transversalis : these, when they reach the edge of the rectus, form the appearance named linea semilunaris ; they then split and enclose the rectus in their duplicature : the whole tendon of the external oblique, with the anterior layer of the internal oblique, passes before the rectus ; and the whole posterior layer of the internal oblique, together with the whole tendon of the transversalis muscle passes behind the rectus, excepting at the lower part ; but, for two or three inches above the pubis, all the tendons go in front of the muscle, and the posterior part of the sheath is consequently deficient, the rectus lying naked on the peritoneum, with only a little cellular tissue interposed.

The two oblique muscles are now to be replaced ; then, making an incision by the side of the linea alba, and thus opening the sheath of the rectus through its whole length, you dissect it back towards the linea semilunaris, and thus lay bare the fibres of the muscle next to be described.

4. Rectus Abdominis.—*Arises,* by a flat tendon, from the fore part of the os pubis : as it ascends, parallel to its fellow, its fleshy belly becomes broader and thinner.

Inserted, by a thin fleshy expansion, into the ensiform

cartilage, and into the cartilages of the three inferior true ribs.

Situation: This pair of muscles are situated on each side of the linea alba, under the tendons of the oblique muscles. The muscle is generally divided by three tendinous intersections; the first is at the umbilicus, the second where it runs over the cartilage of the seventh rib, and the third in the middle between these; and there is commonly a half intersection below the umbilicus. By these intersections, which seldom penetrate the whole substance of the muscle, the rectus is connected firmly to the anterior part of its sheath, forming the LINEÆ TRANSVERSÆ; while it adheres very slightly by loose cellular substance to the posterior layer. Each rectus is contained in a distinct sheath: cut the muscle across midway between the umbilicus and pubis, and, on separating the two portions, the posterior deficiency of the sheath will be observed, marked by a lunated edge. Above, the rectus intermixes with the last slip of the pectoralis major, and it is not unusual to find supernumerary fasciculi ascending still higher.

Use: To compress the fore part of the abdomen, to bend the trunk forwards, or to raise the pelvis.

On each side of the linea alba, and enclosed in the lower part of the sheath of the rectus, is sometimes found a small muscle, named

5. PYRAMIDALIS.—*Origin:* Tendinous and fleshy, of the breadth of an inch, from the os pubis, anterior to the origin of the rectus.

Insertion: By an acute termination, near halfway between the os pubis and umbilicus, into the linea alba and inner edge of the rectus muscle.

Situation: This small triangular pair of muscles lie,

close to the linea alba, between the lower portions of the two recti.

Use : To assist the lower part of the rectus, and to make the linea alba tense.

SECTION II.

DISSECTION OF THE CAVITY OF THE ABDOMEN.

THE abdomen is divided into three regions, each of which is again subdivided.

1. The EPIGASTRIC, or upper region, includes the part covered at the side by the ribs ; its lateral portions are named the right and left HYPOCHONDRIA, and the depression in its middle the *Scrobiculus cordis.*

2. The UMBILICAL, or middle region, is the space immediately under the former ; and it extends below to the anterior superior spines of the ilia. Its sides are called the flanks, or *Lumbar regions.*

3. The HYPOGASTRIC, or lowest region, is subdivided into three parts, one middle, the *Hypogastrium,* and two lateral, named the *Iliac* regions : but immediately above the os pubis, the middle part of this region is called the *Regio Pubis,* and the lateral parts connected with the thighs, are termed the *Inguinal regions.*

Make a longitudinal incision from the scrobiculus cordis to the umbilicus, and from that point an oblique incision on each side towards the anterior spinous process of the os ilium, forming thus three triangular flaps. In doing this, avoid cutting the intestines, by raising the muscles from them after the first puncture.

Before you disturb the viscera, observe the general

situation of those parts which appear on the first opening of the abdomen.

1. The internal surface of the PERITONEUM, smooth, shining, and colourless, covering the parietes of the abdomen, and the surface of all the viscera.

2. In the triangular portion of integument folded down over the pubis, three ligamentous cords project through the peritoneum, two running laterally, and the other in the middle, towards the navel. These are the remains of the two Umbilical Arteries and Urachus.

3. The Epigastric Artery, with its two veins, may be seen through the peritoneum, ascending obliquely upwards and inwards from under Poupart's ligament.

4. The upper edge of the Liver is seen extending from the right hypochondriac region, across the epigastric, into the left hypochondriac region; in it a fissure is seen, into which enters, enclosed in a duplicature of peritoneum, the Ligamentum Teres, which was, in the fœtus, the Umbilical Vein. The fundus of the Gall-bladder, if distended, is sometimes seen projecting from under the edge of the liver.

5. The STOMACH will be found lying in the left hypochondriac region, and upper part of the epigastric; but, if distended, it protrudes into the umbilical region.

6. The GREAT OMENTUM proceeds from the great curvature of the stomach, and stretches down like a flap over the intestines.

7. The GREAT TRANSVERSE ARCH OF THE COLON will be seen projecting through the omentum; it mounts up from the os ilium of the right side, crosses the belly under the edge of the liver, and under the great curvature of the stomach, and descending again upon the left side, sinks under the small intestines, and rests upon the wing of the left os ilium.

8. The SMALL INTESTINES lie convoluted in the lower part of the belly, surrounded by the arch of the colon.

Such is the general appearance on first opening the abdomen; but this will vary somewhat, as one intestine may happen to be more inflated than another, or as the position of the body may have been after death.

It will now be proper to consider the parts more minutely.

1. The PERITONEUM.—Observe how it is reflected from the parietes of the abdomen over all the viscera, so that they may be said to be situated behind or on the outside of it, and it thus forms a bag without an opening, which has been likened to a double night-cap : trace its reflections from side to side, and from above downwards ; you will see that the external coat of every viscus, and all the connecting ligaments, are reflections or continuations of this membrane.

(1.) The FOUR LIGAMENTS of the LIVER are formed by the peritoneum, continued from the inferior surface of the diaphragm and abdominal parietes. *a.* The *middle, suspensory,* or *falciform Ligament,* enclosing in its duplicature the *Ligamentum teres,* and connecting the liver to the diaphragm and linea alba. *b.* The *Coronary ligament,* on the upper and back part of the liver, where that viscus and the diaphragm are in immediate apposition for the space of three or four inches. *c. d.* The *right* and *left lateral Ligaments.*

(2.) The LESSER OMENTUM, or EPIPLOON, or the MESOGASTRION, is formed by two laminæ of peritoneum, passing from the under concave surface of the liver to the lesser curvature of the stomach, and containing in its duplicature the vessels of the liver.

(3.) The GREAT EPIPLOON or OMENTUM. Observe, that the peritoneum, coming from both surfaces of the stomach, and from the spleen, proceeds downwards into the abdomen, and is then reflected back upon itself, till it reaches the transverse arch of the colon, where its laminæ separate to invest that intestine. This reflection is named the Great Omentum; it is a pouch or bag, composed of four laminæ of peritoneum, and the opening into it is by the FORAMEN of WINSLOW. Observe the situation of this semilunar opening; it is on the right side of the abdomen, at the top of the lesser lobe, or lobulus spigelii of the liver; it leads under the little epiploon, under the posterior surface of the stomach, but above the pancreas and colon, into the sac of the omentum. The omentum sometimes reaches to the lower part of the hypogastric region, sometimes not beyond the navel; it contains in its duplicature more or less of adipose substance.

(4.) The MESENTERY. Observe, that the peritoneum, reflected from each side of the vertebræ, proceeds forward, to connect the small intestines loosely to the spine; that it begins opposite to the first lumbar vertebra, crosses obliquely from left to right, and ends halfway between the last lumbar vertebra and the groin. At its commencement, it binds down the extremity of the duodenum; and it terminates where the head of the colon begins. The great circumference which is in contact with the intestines, is very much plaited or folded, and is several yards in length. The mesentery is thus composed of two laminæ of peritoneum, which are easily separable, and between which we always find cellular tissue, fat, lymphatic glands, the trunks and branches of the mesenteric vessels, and the nervous plexuses which accompany them, and also many lacteals and lymphatics, which may sometimes be inflated by the blow-pipe.

(5.) The MESOCOLON is similar to the mesentery,

connecting, in like manner, the colon to the spine, but less loosely : it divides transversely the abdominal cavity into two compartments, one inferior, which is filled by the small intestines ; the other, superior, which contains the stomach, spleen and liver.

2. HEBAR, the LIVER. *Situation* : Partly in the right hypochondrium, which it fills up, reaching as low as the kidney of that side ; partly in the epigastrium, and extending also some way into the left hypochondrium.

Connected by its four ligaments to the inferior surface of the diaphragm, and by the lesser epiploon to the small curvature of the stomach :—The little epiploon should now be removed, to discover the different parts of the liver.

Observe the superior or convex surface adapted to the arch of the diaphragm ; the inferior or concave surface resting on the stomach :- the posterior or thick edge lying against the vertebræ, and the anterior thin margin corresponding to the lower edge of the chest. Observe the three principal lobes of the liver ;—the great or *right lobe*, the smaller *left lobe ;* the *lobulus spigelii*, triangular and projecting, situated behind the lesser epiploon ;— the *longitudinal fissure*, more or less deep, separating the right and left lobe, sometimes converted into a canal by the liver stretching across, and containing the *suspensory ligament* and *ligamentum teres*, and posteriorly the *ductus venosus ;*—the *transverse fissure*, extending along the under surface of the right lobe, and crossing the longitudinal fissure at right angles ;—the *posterior fissure* between the right lobe and lobulus spigelii for the vena cava inferior, which fissure is almost a complete foramen ; —the *depression* in the right lobe for the gall bladder : remark also two smaller lobes or eminences of the liver ; the *lobulus quadratus*, square and little prominent, situated in front of the transverse fissure, at the under surface of the great right lobe, between the gall-bladder and round ligament;—the *lobulus caudatus*, stretching obliquely, from the lobulus spigelii, to the right lobe. Observe the vessels lodged in the transverse fissure, or *portæ* of the liver; the *hepatic artery* on the left side, the *ductus communis choledochus* on the right side, splitting into the *hepatic* and *cystic* ducts, and betwixt, but at the same time behind them, the *sinus* or *trunk* of the *vena portæ*, bifurcating into its right and left branches, which diverge horizontally. These vessels are surrounded by nervous filaments and lymphatic vessels, and they pass along the right edge of the mesogastrion, or lesser omentum, surrounded and connected by adipose and cellular substance:—this edge of the lesser omentum has a thickened feel, it is loose and unattached, and extends from the fissure of the liver to the pylorus, and is called the *Capsule* of *Glisson;* and immediately behind it is the *foramen of Winslow.* Observe that the ligamentum teres, which was the umbilical vein of the fœtus, entered the vena portæ; and that the ductus venosus, now obliterated into a fibrous cord, leaving the vena portæ, passed into one of the venæ cavæ hepaticæ.

3. Vesicula Fellis, the Gall-Bladder. *Situation:* In the right hypochondrium, in a superficial depression on the under surface of the right lobe of the liver, closely united behind to the substance of the liver by cellular tissue, and covered, anteriorly, by the peritoneum passing off from the liver. *Shape*, pyriform; its *position* oblique, so that the *fundus* inclines downwards and to the right, whilst the *body* and *neck* are directed backwards, upwards, and to the left: it rests, below, on the pylorus, duodenum and arch of the colon. The neck is slightly curved, contracting gradually to form the *Ductus*

Cysticus, which is an inch and a half long, and unites with the *Ductus Hepaticus.* This latter duct, of the same length, but larger, springs from the transverse fissure of the liver by a right and left branch. The common canal, or *Ductus Communis Choledochus,* formed by this union of the hepatic and cystic ducts, is four inches long ; it passes behind the right extremity of the pancreas, and perforates obliquely the duodenum, either singly, or united with the pancreatic duct.

4. VENTRICULUS, the STOMACH. *Situation :* In the left hypochondriac and epigastric regions : Connected to part of the inferior surface of the diaphragm, to the concave surface of the liver by the little epiploon, to the spleen by a reflection of peritoneum, and to the arch of the colon by the great omentum. Observe its *Greater Arch* or *Curvature* looking downwards, its *Lesser Curvature* looking upwards ; and its two *lateral* surfaces. In the living body the greater curvature is turned forward, and a little downward ; the lesser arch backward, *i. e.* toward the spine; while one of the lateral convex sides is turned upwards, and the other downwards. Observe the *bulging extremity* on the left side ; the *Cardia* or upper orifice, where the œsophagus enters, and which is half-way between the great extremity and the lesser arch ; the *Pylorus,* or lower orifice, at the end of the small extremity, situated under the liver, and rather to the right side of the spine, feeling hard when touched.

5. The INTESTINES. These form one continuous tube, about six times the length of the body ; but are divided into two portions, differing in their figure, structure, and functions, and distinguished by the names of small and large.

The Small Intestine is divided into *Duodenum, Je-*

junum, and *Ileum;* the large into *Cæcum, Colon,* and *Rectum.*

(1.) SMALL INTESTINE;—comprising the upper four fifths of the Canal.

a, The DUODENUM is broader than any other part of the small intestine, but is short; it takes a turn from the pylorus upwards, and to the right side, passing under the liver and gall-bladder; then, turning upon itself, it descends, passing in front of the right kidney; it is in this space that it receives the pancreatic and gall ducts; thence it crosses before the renal vessels, before the aorta, and upon the upper vertebræ of the loins, firmly bound down by the peritoneum, which covers only its anterior surface; it then ascends from right to left, till it is lost under the root of the mesocolon.

Turning back the colon and omentum, fixing them over the margin of the thorax, and depressing the small intestines towards the pelvis, you find the duodenum coming out from under the mesocolon, but still tied close to the spine; it terminates in the jejunum, exactly where the mesentery begins, and where the intestine becomes loose in the abdominal cavity. The duodenum in this course forms nearly a circle, the root of the mesocolon being the only part lying between its two extremities.

You have now to trace the rest of the small intestine, which lies convoluted in the umbilical and hypogastric regions.

b, The JEJUNUM constitutes the upper two-fifths of the remaining small intestine, and is situated more in the upper part of the abdomen; it is redder, and its coats feel thicker to the touch, from the greater number of the valvulæ conniventes on its inner surface: its diameter exceeds that of the ileum.

c, The ILEUM comprises the lower three-fifths of the

small intestine : it is situated more in the lower part of the abdomen, is of a paler colour, and its coats thinner : it terminates in the great intestine, by opening into the cæcum.

As a general observation it may be said, that the convolutions of the small intestine occupy the middle of the umbilical and hypogastric regions; but their situation varies much, particularly according to the state of the bladder and rectum. The course of the tube, independently of its convolutions, is from the left lumbar region, where the duodenum emerges from under the mesocolon, to the right iliac fossa, where the ileum terminates in the cæcum.

(2.) GREAT INTESTINE.

d, The CÆCUM, or blind gut, is tied down by the peritoneum to the right iliac fossa, which it nearly fills up : sometimes it has a loose peritoneal fold, or *meso-cæcum*. It forms a short rounded pouch; receiving the ileum on its left side, and projecting for an inch and a half below the entrance of the small intestine. On its posterior part there is a little appendage, of the shape of an earth-worm, named *Appendix Cæci Vermiformis*.

e, The COLON is the continuation of the cæcum, which is frequently termed *caput coli ;* it mounts upwards from the cæcum over the anterior surface of the right kidney, to which it is connected by cellular substance, this is called the *ascending portion* of the colon ; it next passes under the gall-bladder, which, after death, tinges it with bile; and, then crossing from right to left, below the stomach, to which it is tied more or less closely by the great epiploon, it forms its GREAT TRANSVERSE ARCH. In its whole course, the great intestine is contracted into cells by its muscular fibres, which are united together, forming longitudinal bands ; and it has some fatty projections attached to its surface, named *Appendices Epiploicæ*. Both these circumstances distinguish the large from the small intestine, which the difference of size does not always, since that varies according to the state of distension. The colon then goes backwards, under the stomach and spleen, into the left hypochondrium; and descending over the left kidney, along the left lumbar region, is again tied down; this is called the *descending portion;* then reaching the left iliac fossa, it is again unconfined, forming a loose and remarkable curvature, which is named the Sigmoid, or Iliac Flexure. After this convolution, the intestine inclines inwards, passing over the brim of the pelvis, and assumes the name of

f, The Rectum.—Drawing aside the intestines, you find the gut continued, over the anterior surface of the sacrum and os coccygis, to the anus, and tied to the former bone by a short reflection of peritoneum; called the *meso-rectum.*

On pulling the stomach towards the right side, you will perceive,

6. The Lien, or Spleen.—*Situation*: In the back part of the left hypochondriac region, between the great extremity of the stomach, and the neighbouring false ribs, under the edge of the diaphragm, and above the left kidney; to all of which it is connected by the peritoneum. It is of an oval figure: its external surface is gently convex: its internal surface irregularly concave, and divided by a *longitudinal fissure*, into which its vessels enter.

7. The Pancreas.—*Situation*: This gland, which was partly seen on removing the little epiploon, is fully exposed, by tearing through the great epiploon between the great curvature of the stomach, and the transverse

arch of the colon. · It lies in the cavity into which the
foramen· of ·Winslow leads : it is of a whitish colour,
and of·an oblong flattened shape, extending from the
fissure of ·the spleen across the spine, under the posterior surface of the stomach, and terminating within the
circle formed by the duodenum, by a broader portion or
head, which has sometimes a detached lobe, or *lesser
pancreas.*

The Pancreas is covered only on its anterior surface
by peritoneum; it lies plunged in a cellular mass, anterior to the aorta, vena portæ and crura of the diaphragm,
with the cœliac trunk and ganglia of the solar plexus
above·it, and the great mesenteric artery descending behind it. The *Pancreatic Duct* runs from left to right
in the substance of the gland, and, emerging from the
great extremity, pierces the duodenum, to open by the
orifice common to it and the ductus choledochus. To
discover the duct, make a horizontal incision, when it
will be distinguished by the whiteness of its coats. It
is of the size of a crow quill.

All the abdominal viscera may now be removed, except the rectum, which, being tied, should be allowed
to remain, for it belongs to the demonstration of the
pelvic viscera and perineum. Or the liver and its vessels, with the pancreas, may be left, and the vessels entering the portæ of the liver traced.

. The peritoneum should then be carefully dissected
from the diaphragm, and from the sides and back part
of the abdomen ; thus the parts, which lie more immediately behind that membrane, may be examined.

8. RENES, the KIDNEYS.—These two bodies are
situated deeply in the posterior part of the abdominal
cavity, on each side of the spinal column, opposite to the
two last dorsal and two first lumbar vertebræ; they lie

in front of the two last ribs, and on the quadratus and psoæ muscles, embedded in soft fat, and covered anteriorly by peritoneum, and by the ascending and descending portions of colon. The right kidney is somewhat lower than the left.

In each kidney, you may observe a lesser arch or concavity, inclined forwards and inwards ; a greater arch or convexity, directed backwards and outwards ;—two lateral surfaces ;—two extremities, the superior of which is nearer to that of the opposite kidney than the inferior. Observe the renal or emulgent artery entering the lesser arch, or sinus ; the vein and ureter passing out ; the venous trunk anterior, behind it the artery, (dividing, like the vein, into four or five branches,) and, posterior to both, the pelvis or commencement of the ureter. Observe the course of the URETER, which is of the size of a common quill ; it descends behind the peritoneum, parallel to the vertebral column, over the psoas muscle; into the pelvis ; it crosses the common iliac artery and vein in its oblique descent, and then runs between the rectum and bladder, which last it enters.

9. The CAPSULÆ RENALES, or *Supra-renal capsules.* —Two glandular bodies, *situated* on the upper extremity of each kidney, of an irregular figure, crescent-like, or somewhat triangular.

By the removal of the peritoneum, several muscles are exposed, at the upper and posterior parts of the abdominal cavity.

One single muscle is situated at the upper part.

The DIAPHRAGM, or MIDRIFT.—This is a broad, thin, muscular septum, placed between the thorax and abdomen, concave towards the latter cavity, and convex towards the thorax, into which its vaulted part ascends, on each side, as high as the fifth or sixth rib. It is fleshy in its circumference, and tendinous in its centre; nearly circular, but somewhat larger transversely. It may be divided into two portions:

1. The *superior*, or *greater* muscle of the diaphragm, or *costal* portion, forms the transverse partition between the chest and abdomen;

Arising, by distinct fleshy fasciculi, (1) *Anteriorly*, from the posterior surface of the ensiform cartilage; (2) *Laterally*, from the inner surfaces of the cartilages of the seventh and of all the false ribs, on each side, its rounded slips intermixing with those of the transversalis muscle: and (3) *Posteriorly*, from the ligamentum arcuatum, which is a ligament extended, somewhat indistinctly, from the extremity of the last rib to the transverse process of the first lumbar vertebra, forming an arch over the psoas and quadratus lumborum muscles. From this circular attachment to the base of the chest, the fibres run in different directions, like radii of a circle, to be

Inserted into the anterior, lateral and posterior borders of the shining, broad, *cordiform tendon*, which occupies the centre of the diaphragm, and which is composed of tendinous fibres, interlaced in different directions.

2. The *Inferior*, or *Lumbar* portion, or *Appendix* of the diaphragm, consists of two fleshy pillars, or *Crura*, placed on the bodies of the vertebræ. Each Crus

Arises from the three or four upper lumbar vertebræ, by four small tendinous feet, which soon unite to form a strong pillar: the left Crus is smaller than the right, and arises from the vertebræ higher up. These two Crura of the diaphragm ascend, leaving a posterior interval between them for the aorta, and form two fleshy bellies, from each of which a fasciculus is detached, which, crossing over, decussates with the opposite one, and then again ascends, (like the letter X,) so that a second or upper opening is formed for the œsophagus. The fleshy fibres of the crura are finally

Inserted into the posterior border of the central tendon.

Situation: The diaphragm is covered on its upper surface by the pleura; below, by the peritoneum, which adheres closely to the central aponeurosis. It separates the thoracic from the abdominal viscera. It has three principal *apertures,* or perforations, for the passage of important organs.

(I.) The *Aortic opening,* or *hiatus,* described above, close upon the spine, which completes it as a fibrous ring. Besides the aorta, this opening gives also a passage to the thoracic duct, and frequently to the vena azygos.

(2.) The *Œsophagean fissure,* or *foramen,* above and to the left side of the aortic opening: it is oval, its sides fleshy, and an inch and a half in extent, giving passage to the œsophagus, with the two nerves of the eighth pair attached to it. These two openings are separated by the decussating fasciculi, above described, of which, that from the right side is always larger, while the left fasciculus is usually anterior.

(3.) The *foramen quadratum,* or tendinous opening for the vena cava, in the right lateral part of the central tendon, to the right and anterior to the œsophageal opening; its margins are prolonged on the cava.

(4.) The splanchnic nerves, and the continued trunk of the great sympathetic, pierce some of the posterior fibres of the crura; and (5.) on each side of the sternum, there is a small fissure, where the peritoneum and pleura are only separated by adipose tissue.

Use.: The diaphragm is one of the chief agents in respiration; when its fibres contract, the muscle descends, chiefly its lateral portions, enlarging the cavity of the chest vertically, and producing inspiration: in expiration, it is relaxed, and re-ascends, resuming its vaulted form, so as to diminish the thoracic cavity. It also acts in coughing, laughing, the hiccough, &c., and assists the other abdominal muscles to expel the urine and fæces.

Four pairs of muscles are situated within the posterior part of the cavity of the abdomen.

1. The Psoas Parvus.—*Arises*, fleshy, from the sides of the last dorsal, and first lumbar vertebræ; it sends off a slender long tendon, which, running along the anterior and inner side of the psoas magnus, is

Inserted, thin and flat, into the brim of the pelvis, at the junction of the os ilium and pubis: it is also connected with the iliac fascia.

Situation: This slender muscle is sometimes wanting; when it exists, it is in front of the psoas magnus.

Use: To assist the psoas magnus in bending the loins forward: in some positions it will bend the pelvis on the loins.

2. The Psoas Magnus.—*Arises*, from the side of the body, and from the transverse process of the last vertebra of the back, and, in the same manner, from all those of the loins, by two ranges of short, flat, triangular slips; —the nerves of the lumbar plexus coming out between these two portions of the muscle. It forms a thick fleshy belly, which runs down over the brim of the pelvis, beneath Poupart's ligament, into the thigh.

Inserted, tendinous, into the trochanter minor of the os femoris, and, fleshy, into that bone immediately below the trochanter.

Situation: Close to the sides of the lumbar vertebræ, with the psoas parvus on its anterior part; then, descending along the margin of the superior aperture of the pelvis, with the iliacus internus on its outer side. At its origin, it has some connexion with the diaphragm and ligamentum arcuatum, and lies in front of the quadratus lumborum.

Use : To bend the thigh forwards, and roll it outwards; or, when the inferior extremity is fixed, to assist in bending the body.

3. The ILIACUS INTERNUS.—*Arises,* fleshy, from the transverse process of the last vertebra of the loins, from all the inner margin of the spine of the os ilium, from the edge of that bone between its anterior superior spinous process and the acetabulum, and from all its hollow part between the spine and the linea innominata. Its fibres descend under the outer half of Poupart's ligament, and join the tendon of the psoas magnus.

Inserted, with the psoas magnus.

Situation : It fills up the internal concave surface of the os ilium, and is situated on the outer side of the psoas magnus. This muscle, as well as the lower part of the psoas, is covered by a strong fascia, which is inserted into the crista of the ilium, the border of the pelvis, and into the crural arch, and is called the *Fascia Iliaca.*

Use : To assist in bending the thigh, and in bringing it directly forwards.

The insertion of the two last-described muscles cannot be seen till the thigh is dissected, when it will be found to lie between the vastus internus and the pectineus. The common tendon of the two muscles passes over the capsule of the hip-joint, and there is interposed a large bursa mucosa.

4. The QUADRATUS LUMBORUM.—*Arises,* tendinous and fleshy, from the posterior third part of the spine of the os ilium, and from the ilio-lumbar ligament.

Inserted into the transverse processes of all the vertebræ of the loins, into the posterior half of the last rib, and, by a small tendon, into the side of the last vertebra of the back.

Situation: By the sides of the lumbar vertebræ, between the back part of the crista ilii and twelfth rib, posterior to the psoas muscle and kidney. It is enclosed between the layers of the posterior tendon of the transversalis muscle, and forms a portion of the abdominal walls, having the fleshy mass of the sacro-lumbalis and longissimus dorsi posterior to it.

Use: To bend the loins to one side, to pull down the last rib, and, when both muscles act, to bend the loins forwards.

The Fascia Iliaca may be further described as a thin but dense aponeurosis, which, extending from the crista ilii, completely covers the iliacus internus and psoas muscles, adheres firmly to the superior margin of the pelvis, and is then continued downwards into the cavity to line its parietes. Anteriorly, this iliac fascia descends to Poupart's ligament; and it is firmly attached to that ligament, from the spine of the ilium as far down as where the iliac vessels pass beneath the ligament into the thigh; and at this point the fascia has a distinct falciform edge. Here the iliac fascia appears to unfold itself, being in part firmly united to Poupart's ligament, while, behind, it is continued over the common mass of the psoas and iliacus into the thigh. This posterior continuation of the fascia passes behind the iliac vessels, over the brim of the pelvis, forming the floor on which the vessels lie; is closely connected with, and assists to form, their sheath; and becomes continuous, behind these vessels, with the fascia lata of the thigh.

In this dissection of the muscles going down from the

pelvis into the thigh, the student should examine, from within the abdomen, the CRURAL ARCH; the passage of the *external iliac* vessels beneath it; and the manner in which the abdominal cavity is shut up at this part, so that the bowels are prevented from descending under the arch. This demonstration is more advantageously made before dissecting these muscles, and may be conducted in the following manner.

Strip the peritoneum from the inside of the crural arch and from the iliac fossa; and, having removed loose cellular tissue and some lymphatic glands, look downwards under Poupart's ligament, the firm border of which is readily felt: you perceive that the large space or opening which is observed in the skeleton above the anterior border of the os ilium and os pubis, is completely filled up, and that it is bounded above by the line of Poupart's ligament : the outer half or rather more of this space is occupied by the psoas and iliacus muscles, which are seen proceeding from the vertebræ and iliac fossa, invested by the fascia iliaca, and passing down into the thigh : the inner part exhibits the appearance of an *oval* or *triangular space*, into which the *external iliac artery* and *vein*, having descended along the edge of the psoas, are seen entering. This space or passage of the iliac vessels is bounded, posteriorly, by the os pubis ; on its inner side, by *Gimbernat's ligament*, (which is readily felt and exposed ;) on the outer side by the united mass of the psoas and iliacus : the fascia iliaca also closes the outer part by its close adhesion to Poupart's ligament, which ligament is stretched across and bounds the passage anteriorly. The *great artery* and *vein* are not lying loose in this oval space, but are closely connected with the surrounding parts : behind they are tied down by a thin membrane to the iliac fascia, while, anteriorly, tendinous fibres, which may be considered as prolonged from the fascia transversalis, pass down from Poupart's ligament to be connected with the fore-part of the sheath of the vessels. It will now be seen that these two fasciæ, (the *fasciæ iliaca and transversalis,*) are united into one continued aponeurotic sheet along the line of Poupart's ligament; but that at the point, where the iliac vessels are seen passing beneath the ligament into the thigh, the two fasciæ separate in a funnel-like form, and become closely adherent to the anterior and posterior parts of the sheath of the vessels :—This sheath, as it passes under the crural arch, becomes flattened, and is frequently termed the *Crural* or *Femoral Sheath.* The junction of the two fasciæ along Poupart's ligament is marked by a white line :— on the pubic side of the vessels, the fascia transversalis adheres to the ligamentous covering of the crest of the pubis, becoming distinctly continuous with the posterior lamina of *G*imbernat's ligament, while the iliac fascia, passing behind the vessels *, is also fixed to the whole length of the pubic crest, again united to the transversalis fascia, and, below, continued with the pubic or pectineal portion of the fascia lata.

* This passage, by which the iliac vessels are transmitted into the thigh, is now frequently called the *Crural Canal,* the term *Crural Ring* being usually limited to the space between the great vein and Gimbernat's ligament, by which Crural hernia escapes. These two openings are separated by a thin but evident septum. The upper or abdominal orifice of the common passage has an arched form, which is chiefly owing to some stronger transverse fibres of the transversalis fascia : these have been described and delineated by Hesselbach, as the *internal inguinal ligament,* stretching over the great vessels in a plane posterior to the *external inguinal* or *Poupart's* ligament. The lower or femoral outlet of this passage is the *Saphenic opening* in the groin.

There is, however, one weak point beneath the line of the crural arch; pass your finger close to the edge of Gimbernat's ligament; you will find a space between that ligament and the iliac vessels, (more immediately the great vein,) which is only occupied by cellular tissue and one or two lymphatic glands: this is the *Crural Ring*, and it leads into the thigh on the pectineus muscle, as was seen in the dissection of the groin.

The *Anterior Crural Nerve* does not join the outer side of the external iliac artery, until it has fairly passed into the thigh : within the pelvis, the nerve is found, under the iliac fascia, between the contiguous margins of the psoas and iliacus. Some cutaneous nerves from the lumbar plexus * pass with the vessels under the crural arch.

SECTION III.

OF THE VESSELS AND NERVES SITUATED BEHIND THE PERITONEUM.

I. THE ARTERIES, VIZ. THE AORTA ABDOMINALIS, AND ITS BRANCHES †.

THE AORTA passes from the thorax into the abdomen, between the crura of the diaphragm, close upon the spine. It then takes the name of the abdominal aorta, and descends on the fore-part of the vertebræ, lying not

---

* *The genito-crural branch.*

† In the description of the blood-vessels, the ramifications of the principal trunks are enumerated; but the student must remember that these can be seen only when injected, and when the subject is dissected for the express purpose of tracing the arteries. In an ordinary dissection, the trunks only can be demonstrated.

exactly in the middle, but rather inclined to the left side. On the fourth lumbar vertebra, it bifurcates into the two PRIMITIVE or COMMON ILIAC Arteries.

The Aorta is at first concealed by the fleshy pillars of the diaphragm; the Vena Cava is to the right, but separated by the corresponding pillar of the diaphragm; lower down, the great vein is close on the right side of the artery, with only a cellular layer interposed. Anteriorly and on its left side, the aorta is covered by peritoneum; it passes successively behind the liver and stomach, is crossed by the pancreas and lower portion of the duodenum, and, further down, is placed behind the mesentery and great mass of intestine. In this course, it gives off, from its fore part, three single arterial trunks, and, from its sides, several arteries, which come off in pairs.

1. The two PHRENIC Arteries arise from the Aorta, before it has fairly entered into the abdomen, and ascend immediately on the pillars of the diaphragm, to ramify on the broad upper part of the muscle: sometimes they come off in a single trunk, or one artery arises from the coeliac.

2. The COELIAC ARTERY comes off at the point, where the aorta has scarcely extricated itself from the diaphragm: it is a single, large, but short trunk, coming off at right angles, situated between the inferior surface of the liver, and the small curvature of the stomach; and surrounded by the meshes of the semilunar ganglion. It divides at once into three branches, which diverge, as from a centre, and the trunk is called the *Axis Arteriæ Cœliacæ.*

(1.) A. CORONARIA VENTRICULI, the smallest of the three, passes from the axis towards the left side, and, arriving at the cardiac orifice of the stomach, attaches itself to that organ, and is then continued along the lesser curvature from left to right, to inosculate with the pyloric branch of the hepatic artery. Its branches are,

(*a,*) *Arteries,* encircling the cardiac orifice, and others ascending on the œsophagus.

(*b,*) *Gastric Branches,* to the anterior and posterior surfaces of the stomach. Sometimes a large branch is sent to the liver.

(2.) ARTERIA SPLENICA, the largest branch of the cœliac, goes directly to the left side, passes under the stomach, and, running tortuously along the upper border of the pancreas, enters the concave surface of the spleen in several branches; but, previously it gives off,

(*a,*) *A. Pancreaticæ Pärvæ* to the pancreas.

(*b,*) *Vasa Brevia* to the bulging extremity of the stomach, passing in the fold of peritoneum, which is reflected from the spleen to the stomach.

(*c,*) *A. Gastro-Epiploica Sinistra,* which comes off from the splenic, while passing under the stomach; and attaching itself to the bulging end, runs along the greater curvature from left to right, inosculating with the gastro-epiploica dextra.

(3.) ARTERIA HEPATICA runs in a direction opposite to the splenic, passing upwards and to the right side, to reach the transverse fissure of the liver; but not more than half of its blood goes to that organ. It sends off

(*a,*) *A. Pylorica* which descends to reach the pyloric end of the stomach, and turns along the lesser curvature to inosculate with the coronary artery.

(*b,*) *A. Gastro-Epiploica Dextra,* or *Gastro-Duodenalis,* which is a large artery, passes under the pylorus, to reach the great curvature of the stomach, along which it runs, inosculating with the gastro-epiploica sinistra, and sending branches upwards to the sto-, mach, and downwards to the omentum : it also sup-, plies the upper part of the duodenum, and sends off a considerable branch to the right extremity of the pancreas.

The hepatic artery then divides into its right and left branches, which are distributed to the corresponding lobes of the liver ; the right branch giving off the *Cystic Artery* to the gall-bladder.

3. The SUPERIOR MESENTERIC ARTERY is another single trunk, which leaves the aorta about half an inch lower than the cœliac artery, nearly equalling it in size, and descends behind the pancreas; it comes out from under the mesocolon, and stretches over the duodenum ; then entering the root of the mesentery, it passes downwards between the two laminæ, gradually incurvating from left to right, and approaching the intestine. The superior mesenteric gives, at its origin, some small arteries to the pancreas and duodenum, but its chief branches come off in two sets :

(1.) From the right side or concavity of the arch, branches are sent off to the right lateral part of the great intestine : these are usually three in number, coming off at the distance of an inch from each other, and they form arches of anastomosis in the mesocolon.

(*a,*) *A. Ileo-Colica,* or the *Cæcal Artery,* is the lowest branch ; it runs down to the cæcum and last turns of the ileum, inosculating with the last branches of the superior mesenteric sent to the small intestines, and sending a branch upwards along the great intestine, to communicate with the next branch.

(*b,*) *A. Colica Dextra,* passing transversely to the right or ascending colon, and subdividing into two branches, to communicate with the last and following artery.

(c,) *A. Colica Media.* This branch goes directly upwards from the trunk of the superior mesenteric, as it comes out from under the mesocolon : it passes towards the transverse arch of the colon, soon dividing into two branches, of which one communicates with the colica dextra, while the left branch arches in the opposite direction, and joins the colica sinistra, which is a branch of the inferior mesenteric. These anastomosing arches of the right colic arteries send off branches, which pass to the great intestine. Sometimes there are only two arteries, viz. *ileo-colica,* and *colica dextra.*

(2.) The left side or convexity of the arch of the superior mesenteric sends off from *sixteen* to *twenty arteries,* which are the *arteries* of the *small intestine.* These, after descending for some distance between the laminæ of mesentery, bifurcate into secondary branches, which unite and form arches, and from these again more numerous ramifications pass off, forming three or four successive tiers of arches, gradually decreasing in size, which ultimately distribute their branches to the jejunum and ileum.

4. The RENAL or EMULGENT ARTERIES are two in number. Each artery arising, below the superior mesenteric, from the side of the aorta, passes to the kidney, and after giving twigs to the renal capsule and adipose membrane, enters the lesser arch of the kidney. The right artery is longer than the left, and passes behind the vena cava.

The renal capsule on each side often has a distinct artery from the aorta; the *capsular arteries,* which also give twigs to the crura of the diaphragm and fat of the kidneys.

5. The SPERMATIC ARTERIES are also two; they come

off about an inch below the emulgent, from the fore-part
of the aorta. These slender, tortuous, arteries descend
on the sides of the spinal column, before the psoæ
muscles and ureters, and behind the peritoneum; and
the right artery passes before the vena cava inferior:
they are soon joined by the spermatic veins and some
small nerves; the artery on each side, in the male, then
arrives at the internal ring, and, joined by the vas deferens, passes through the inguinal canal, to enter the
upper part of the testicle in five or six branches. In the
female, the spermatic artery descends into the pelvis, and
passing in the fold of the broad ligament, supplies the
ovarium and fundus uteri.

6. The INFERIOR MESENTERIC is a single trunk, which
comes off rather from the left side of the aorta, about an
inch above its bifurcation; it passes to the left side of
the abdomen, behind the peritoneum, and is then engaged between the laminæ of the iliac mesocolon; it is
destined to the left portion of the colon and to the rectum. Its branches are the *A. Colicæ Sinistræ* two or
three arteries, which sometimes come off by a common
trunk; these anastomose with one another, and, by an
ascending branch, with the colica media, and supply the
left portion of the great arch of the colon and the sigmoid flexure.

The trunk of the inferior mesenteric then descends
upon the back part of the rectum, on which it ramifies
largely, and is termed the *A. Hæmorrhoidalis Superior*
or *Interna.*

7. The LUMBAR ARTERIES are four or five small arteries on each side, which arise from the back part of the
aorta; they pass transversely outwards on the bodies of
the vertebræ, being covered by the psoæ muscles. At
the base of the transverse processes, each lumbar artery

divides into a *posterior* branch, which supplies the lumbar muscles, and sends a twig to the spinal canal; and an *anterior* branch, which passes forwards to ramify in the broad abdominal muscles.

8. A. Sacra Media is a single artery, which arises from the back part of the aorta at its bifurcation, and descends along the anterior surface of the sacrum, communicating with the lateral sacral arteries.

At the fourth lumbar vertebra, the Aorta bifurcates into the two Primitive or Common Iliacs.

The Common Iliac trunks, resulting from this bifurcation of the aorta, diverge from each other at an acute angle, and pass obliquely outwards along the inner edge of the psoas muscle. The right artery is somewhat longer than the left. After a course of two or three inches, and opposite the sacro-iliac symphysis, each of the Primitive Iliacs subdivides into,

(1.) The Internal Iliac, or Hypogastric, which passes down into the pelvis.

(2.) The External Iliac, which, following the direction of the psoas muscle, passes under Poupart's ligament, and becomes the inguinal artery.

Many lymphatic glands are observed around the iliac vessels, and along the line of the aorta and vena cava.

II. VEINS.

The Inferior, or Ascending Vena Cava is formed by the junction of the two common iliac veins, on the right side of the fourth lumbar vertebra. The great vein, thus formed, ascends along the vertebræ, on the right side of the Aorta. It lies close to the artery, until the two vessels approach the diaphragm, when the Cava, inclining forwards and to the right side, is lodged in the posterior fissure of the liver, being nearly surrounded by that viscus. It then passes through the opening in the tendon of the diaphragm, the great artery and vein being here separated by the right crus of the muscle.

. In this course the Cava receives the following veins, which resemble their corresponding arteries, (1.) The *Lumbar veins*. (2.) The *Emulgent* or *Renal Veins*, of which the left is the longest, as it crosses over the fore-part of the aorta. (3.) The *Right Spermatic Vein ;*— the *left* enters the left renal vein. (4.) *Veins* from the renal capsules. And where it passes through the fissure of the liver, it receives the three *Cavæ hepaticæ*, and also in general two **Diaphragmatic Veins**. The Cava then enters the chest, and terminates immediately in the right auricle of the heart.

The COMMON ILIAC VEIN of each side is formed by the union of its two branches, the EXTERNAL and INTERNAL ILIAC VEINS, which accompany the arteries of the same name. Of these, the EXTERNAL ILIAC (which is a continuation of the femoral vein) ascends from Poupart's ligament on the inner side of its corresponding artery, receiving the circumflex-iliac and epigastric veins, and, near the sacro-iliac symphysis, is joined by the INTERNAL or HYPOGASTRIC Vein from the pelvis. The COMMON ILIAC VEIN of each side lies on the inside of its artery, and somewhat posterior ; hence both veins cross behind the right iliac artery, to unite and form the vena cava, on the fore-part of the lumbar vertebræ *.

The veins which return the blood from all the organs contained in the cavity of the abdomen, excepting the kidneys and bladder, and the uterus in the female, do

---

* It may be here proper to observe, that generally a great vein accompanies every great artery ; but when the ramifications become small, each artery is attended by two veins.

not join the vena cava, but unite to form a trunk, called the *Vena Portæ.*

The VENA PORTÆ is formed by the union of two principal branches :

1. The *Splenic Vein*, coming transversely from the spleen, and also receiving the *Inferior Mesenteric* Vein, which leaves its artery, and ascends behind the peritoneum along the left lumbar region.

2. The *Superior Mesenteric Vein*, which passes upwards from the mesentery, where its branches resemble those of the corresponding artery. A third, much smaller vein, the *Coronaria Ventriculi*, also joins the Vena Portæ or its splenic branch, behind the pyloric extremity of the stomach.

The Vena Portæ is found as a trunk, resulting from the union of the two great veins, behind the pancreas; it ascends obliquely to the right side, passing behind the duodenum, towards the transverse fissure of the liver, and is from three to four inches in length. It is enclosed in the lesser omentum, behind the hepatic artery and biliary ducts, and, on reaching the right extremity of the transverse fissure, bifurcates into two branches, which diverge, and ramify anew, (like arteries,) through the corresponding sides of the liver, and the blood is returned into the Vena Cava by the Cavæ hepaticæ.

The *Vena Azygos* is also seen in the abdomen; it is formed of some small veins from the loins, and generally perforates the right crus of the diaphragm, or passes between the crura of the diaphragm by the same opening as the aorta and thoracic duct.

III. NERVES.

1. The eighth pair, or *Nervus Vagus*, is seen descending on each side of the œsophagus, and passing on

it from the thorax into the abdomen. The *right* nerve, which is larger than the left, is placed on the right and posterior part of the œsophagus, and subdivides into numerous filaments, which form around the cardia a considerable plexus, and ramify on the posterior surface of the stomach, joining the filaments of the great sympathetic nerve to form the *coronary* or *stomachic* plexus. The *left* nerve lies on the fore part of the œsophagus, and is distributed to the anterior surface of the stomach, communicating with the right nerve: some branches also pass from the two nerves to the hepatic, splenic, and great solar plexus.

2. The SPLANCHNIC NERVE, or *Anterior Intercostal*, which is formed by filaments from the ganglia of the great sympathetic nerve in the thorax, enters the abdomen between the fibres of the crus of the diaphragm, close upon the vertebræ, or sometimes by the aortic opening; here the nerve, on each side, terminates in the SEMILUNAR GANGLION, by the side of the cœliac artery. The two semilunar ganglia are of large size, resting on the crura of the diaphragm and aorta, and from each ganglion branches are sent across, which communicate intimately together, and form, round the root of the cœliac artery, a very intricate plexus, containing several ganglia of various sizes, and called the SOLAR PLEXUS. Nerves pass from this great plexiform centre, accompanying the divisions of the abdominal aorta, to the various viscera of the abdomen; in a common dissection these nerves cannot be clearly demonstrated, as they lie very close on the respective arteries, and are surrounded by much condensed cellular substance; they form the coronary or stomachic, the hepatic, splenic, superior and inferior mesenteric, renal, spermatic, and hypogastric plexuses. There is also a *lesser Splanchnic nerve*, which pierces

the crus of the diaphragm external to the larger nerve, communicating with it by filaments, and finally terminating in the renal plexus.

3. The continued trunk of the *great sympathetic nerve,* or the *posterior intercostal,* also enters the abdomen, close to the spine, behind the posterior fibres of the diaphragm, to run down on the sides of the lumbar vertebræ, and along the upper edge of the psoas magnus. It forms small ganglia, four or five in number, opposite the bodies of the vertebræ, and, descending into the pelvis, produces four sacral ganglia on the anterior surface of the sacrum, terminating finally on the extremity of the os coccygis, by an arch-like union with the nerve of the opposite side, in a ganglion named *Ganglion Impar.* In this course, it communicates with the lumbar and sacral nerves, and abdominal plexuses.

The THORACIC DUCT may be observed passing from the abdomen into the thorax, between the aorta and the right crus of the diaphragm. It is larger here than in its subsequent course, and the dilated portion is called *receptaculum chyli,* as the trunk of the lacteals pours into it the chyle at this point.

SECTION IV.

OF THE STRUCTURE OF THE ABDOMINAL VISCERA.

MANY particulars respecting the structure of the viscera, may be usefully studied without further preparation of the several parts.

The *Liver ;*—Observe, 1. Its *peritoneal* coat, or tunic. 2. Its *Cellular* tunic, placed under the peritoneal, and also extending into the substance of the viscus, where it is seen forming sheaths for the vessels, and is often called *G*lisson's Capsule. 3. The *Parenchyma*, or substance of the liver, which is porous and brittle, and, when ruptured, has a granulated appearance : the minute granulated bodies appear to be united by cellular tissue, and are supposed to be formed by the ultimate ramifications of the vessels of the liver, and of the hepatic duct. 4. The lower surface of the liver, with its fissures and the vessels in them, may also be examined.

The *Gall-Bladder*:—Observe, 1. Its *Peritoneal* coat, which only covers it in part. 2. The *Cellular* tunic, which is only evident where the gall-bladder is united to the fissure in the liver. 3. The *Mucous* or *Internal* coat, which is tinged with bile after death, and has a rugose or villous appearance, with two or three minute valvular folds at the neck of the gall-bladder. Examine the *Hepatic* and *Cystic* duct, uniting and opening into the duodenum.

The *Stomach*:— Observe its three coats, united by intermediate cellular tissue. 1. Its external *Peritoneal* coats. 2. The *Muscular* coat, the fibres of which are indistinct, except the longitudinal ones, which may be traced from the œsophagus : they are always white ; some semicircular fibres are also observable about the middle of the stomach. 3. The *internal, villous* or *mucous* coat : invert the stomach and wash it, and observe its inner surface, soft and whitish, distinctly villous, besmeared with a viscous fluid, and, when the stomach is empty, thrown into rugæ : the villous coat is connected to the muscular by dense cellular tissue, which has been called the *Nervous coat.* 4. The *Cardiac* orifice, surrounded by the coronary artery and veins and nerves of the eighth pair ; a line of separation is evident on the mucous coat, where the œsophagus terminates.

5. The *Pyloric* orifice; observe the *Valve of the Pylo-rus*, a broad and flattened circular projection, formed by a fold of the muscular and villous coats, which contain between them a firm fibrous annular substance; its inner edge is loose and floating.

The *Intestines*.

(a) The *Small Intestine*. Observe three coats, as in the stomach: 1. *External* or *Peritoneal* coat. 2. *Muscular* coat, consisting of two sets of fibres: *longitudinal*, which are external and indistinct; *circular* or *transverse*, which are found within the former: 3. *Villous* or *mucous* coat, seen on inverting the intestine, red, soft, and villous, and having numerous transverse duplicatures, called *valvulæ conniventes;* in the cellular texture, which unites the muscular and villous coats, (termed the *Nervous* coat,) we meet with numerous mucous glands. The muscular coat of the duodenum is most apparent; in the jejunum the fibres are less evident: both these intestines have numerous valvulæ conniventes; the coats of the ileum are thinner, and its valvulæ fewer and smaller. Remark, within the duodenum, a small eminence, or papilla, pierced by the united, or separate orifices of the ductus communis choledochus and pancreatic duct.

(b) The *Large Intestine*. This has also three coats: 1. The external, or *Peritoneal*. 2. *Muscular;* the longitudinal fibres are collected into three bands, which contract the intestine into *sacculi* or pouches: the circular are disposed as in the small intestine; this *sacculated* appearance commences at the cæcum, and extends to the sigmoid flexure. 3. The *Internal* or *mucous* coat is pale, and much less villous: the large intestines have numerous mucous glands.

Observe that the *Cæcum* is only partially invested by peritoneum : the *Appendix*, or *Processus Vermiformis*, opens into the cæcum, and terminates by a blind extremity. Observe the *Valve* of the *Colon*, or *Ileo-colic Valve*, of an elliptic form, projecting into the cavity of the large intestine, with two lips, which are united at their extremities, formed by a folding of the mucous coat and some muscular fibres. The *Rectum* will be examined with the pelvic viscera : it has no sacculated appearance, but its longitudinal fibres are very distinct, and are spread uniformly over the surface of the intestine ; the circular fibres are also distinct at its lower extremity ; the mucous coat resembles that of the rest of the intestinal canal, but is thicker, more red, with longitudinal folds, and, at the lower part of the rectum, is attached but loosely to the muscular coat : in the dilatation, or cul de sac above the anus, numerous orifices of mucous follicles are seen.

The *Spleen.* Remark 1. Its *Peritoneal* coat. 2. A *Fibrous coat*, adhering closely to the peritoneal coat, and sending processes into the parenchyma. 3. The *Parenchyma*, or substance, soft, and spongy, easily broken down, of a reddish or leaden colour, and apparently consisting of numerous cells, which are generally loaded with blood.

The *Pancreas* has no distinct coat, but lies surrounded by cellular tissue, and is only partially covered by peritoneum ; it is a conglomerate gland, resembling in its structure the salivary glands ; its *parenchyma* is firm and resisting, and appears composed of granulated lobules, which are united by cellular tissue : from these lobules the minute divisions of the *excretory duct* take their origin : the duct is of the same texture as the salivary ducts.

The *Kidneys* are to be examined by making a section. Observe 1. That the kidney has a distinct *membranous* or *fibrous* coat, which invests it closely, and enters into its sinus. 2. The *Parenchyma* is composed of two distinct substances. *(a,)* The external, or *Cortical,* is two or three lines in thickness, of a yellowish red colour, granular, and highly vascular, sending elongations or partitions inwardly, which divide *(b,)* the internal, or *tubular* substance: this is of a pale red colour, formed of several conical bodies, which are surrounded at their base by the cortical substance, and terminate in rounded points, termed *Mamillæ* or *processus Mamillares;* these project into the inner concavity or *Sinus:* this tubular part has a striated appearance, consisting of ducts called the *tubuli uriniferi,* which terminate, by open mouths, in the papillæ. 3. The mamillary processes are surrounded by membranous sacs, called *calyces* or *infundibula,* which unite in forming a membranous reservoir, the *pelvis* of the kidney: the pelvis is situated in the sinus or fissure of the kidney, behind the renal vein and artery; it is of a compressed oval form, and shortly contracts and forms the ureter. The *Ureter* is composed of two coats; the *external* is cellular or fibrous; the *internal* is mucous, and continuous with the mucous lining of the bladder.

The *Renal Capsules* are usually of a yellowish brown colour: in the fœtus, they are large, and their substance lobulated; in the adult they are diminished in size, and have an indistinct oblong cavity within, containing a dark-coloured fluid. In aged subjects, they are found shrunk, or even disappear.

CHAPTER II.

DISSECTION OF THE PERINEUM, AND OF THE MALE ORGANS OF GENERATION *.

The muscles and vessels to be demonstrated, lie deep amongst much loose cellular substance; and unless great caution is used, important parts will be removed, while the student supposes he is only clearing away cellular substance:—the rectum having been cleansed, a little baked hair or a sponge may be introduced into its extremity; or a cork, with a loop attached to it, may be made use of, and the mouth of the gut tied upon it; this is for the purpose of making the anus, and lower part of the intestine, slightly prominent, as a guide to the knife, in clearing the fibres of the sphincter and perineal muscles : the dissection will also be facilitated by introducing a staff into the bladder, to mark out the situation of the urethra; the subject should be placed in the same position as for the lateral operation of lithotomy: and the bladder may be moderately inflated through the ureter.

* This dissection will be more complete, if the pelvis and lower extremities are injected; for thus the important branches of the pudic artery will be more easily traced.

SECTION I.

The muscles of the perineum consist of five pairs, and a single muscle:

ERECTOR PENIS,
ACCELERATOR URINÆ,
TRANSVERSUS PERINEI, } on each side,
LEVATOR ANI,
COCCYGEUS,

SPHINCTER ANI, a single muscle.

Both sides of the perineum should be dissected at the same time. Lift up the scrotum, and make an incision, on each side, in the direction of the rami of the os pubis and ischium, and dissect off the skin, reflecting it inwards to the *Raphe* or line, which, running along the skin from the scrotum to the anus, marks the place where the opposite muscles meet: extend your dissection along the tubera ischii, and to the os coccygis behind.

The integuments being cautiously removed, we find, immediately beneath them, a thin but firm fascia, connected with the scrotum anteriorly, with the fascia of the thighs laterally, and, behind, intermixed with the fat and loose cellular membrane about the tuber ischii and margin of the anus: this is the *Superficial Fascia of the Perineum;* in emaciated subjects, it may be demonstrated as a firm aponeurotic layer; but where there is much fat, the tendinous fibres are dispersed in the layers of adipose tissue.

The muscles are now to be exposed, for which purpose the superficial fascia is to be removed; in prosecuting this dissection, recollect that the erector muscle, covering the crus of the penis, arises from the tuber ischii, and ascends on the inside of the ramus of that bone :—that the transversus perinei arises from nearly the same point, and crosses the perineum, lying often at a considerable depth in the adipose substance. The tuber ischii then becomes a proper place for the commencement of the dissection; carefully tracing the muscles arising from that point, remove all the cellular substance, situated in the perineum, while the muscular fibres are left untouched. The appearance of these muscles will vary in different subjects. In those who have died weak and emaciated, the fibres will be pale, and not very evident, while in strong muscular men, who have expired suddenly, they will be very distinct. This dissection is to be continued, till all the parts between the tuberosities of the ischia on each side, and between the pubis before, and the tip of the os coccygis behind, are fairly brought into view. Observe,

On each side, the *Erector Penis* covering the crus of the penis.

In the middle, the *Accelerator Urinæ,* with its fellow, embracing the bulb and lower part of the corpus spongiosum of the urethra.

Lower down, the *Transversus Perinei,* crossing the perineum transversely.

Behind, the *Sphincter Ani,* encircling the anus : and a large quantity of adipose and cellular tissue, in the space on each side, between the sphincter ani, tuber ischii, and edge of the gluteus maximus, which latter muscle is seen stretching from the coccyx and sacrum to the thigh, and lapping over the tuber ischii and great sacro-sciatic ligament.

1. The ERECTOR PENIS—*Arises,* tendinous and fleshy, from the inner side of the tuber ischii ; its fleshy fibres proceed upwards over the crus of the penis, adhering to

the outer and inner edges of the ascending ramus of the ischium, and descending ramus of the pubis;—but before the two crura meet to form the body of the penis, it ends in a flat tendon, which is lost in the strong tendinous membrane that covers the corpus cavernosum.

Situation: This muscle covers all the surface of the crus penis that is not in contact with bone; it arises on each side of the attachment of the crus to the bone.

Use: To draw the crus penis downwards and backwards.

2. The ACCELERATOR URINÆ—*Arises*, by a thin tendinous expansion, from the descending ramus of the pubis, and from the ascending ramus of the ischium, nearly as far down as the tuber;—this origin lies under the crus of the penis, and the fleshy fibres are seen coming out from the angle between the crus and the corpus spongiosum urethræ; they proceed obliquely downwards and backwards, embrace the bulb and lower part of the corpus spongiosum, and are

Inserted into a white tendinous line, on the middle of the bulb and corpus spongiosum of the urethra, joining there with the muscle of the opposite side. The lowermost fibres run nearly transversely, while the superior fibres are very oblique.

Situation: Immediately under the skin and superficial fascia; the posterior surface of this pair of muscles covers the bulb and part of the corpus spongiosum of the urethra.

Use: To drive the urine and semen forwards, by compressing the lower part of the urethra.

3. The TRANSVERSUS PERINEI—*Arises* from the tough fatty membrane that covers the inside of the tuber ischii, immediately behind the attachment of the erector penis: thence its fibres run transversely inwards.

Inserted into the central point of union, where the sphincter ani touches the accelerator urinæ, and where a kind of tendinous projection is formed, common to the five muscles.

Use: To fix the central point of attachment of the perineal muscles, and to support the anus and perineum generally. It may tend to dilate the bulb of the urethra.

There is sometimes another anterior slip of fibres, the Transversus Perinei Alter, which crosses more obliquely, and is inserted into the posterior part of the bulb of the urethra. There is considerable variety in these transverse muscles, and their fibres are often indistinct. In some subjects, these transverse muscular fibres seem united to the lower border of the triangular ligament.

4. The Sphincter Ani consists of two semicircular planes, which run round the extremity of the rectum, passing nearly as far out as the tuber ischii; the fibres on each side decussate where they meet, and are

Inserted, behind, into the apex of the os coccygis, generally by an elastic ligament or tendon of some length: before, into the tendinous point, common to this muscle, and to the acceleratores urinæ, and transversi perinei. This tendinous point is worthy of remark, consisting in part of an elastic ligamentous substance.

Situation: The external fibres of this muscle, which are immediately beneath the skin, are generally pale and intermixed with adipose substance, and with a few fibres of the superficial fascia, which is weak and indistinct about the anus. These fibres have been occasionally described as a distinct muscle, the *External* or *Cutaneous Sphincter.* The deeper seated part of the muscle is in contact with, and laps over, the lower part of the levator ani.

Use: To close the anus. It is in a state of constant contraction, independently of the will.

More deeply seated than the muscles now described, we see some of the fibres of

The LEVATOR ANI. To expose this muscle, the adipose and cellular substance must be freely removed from the side of the rectum, laying bare the space between the transversus perinei, and the edge of the gluteus maximus; the muscle cannot, however, be seen completely, until the side of the pelvis is removed.

It *arises*, fleshy, from the lower and posterior part of the symphysis pubis, and from the body of that bone above the obturator foramen; thin and tendinous, from the broad fascia covering the obturator internus, and, again fleshy, from the inner surface of the ischium and its spinous process. From this semicircular origin, the fibres run downwards, in a radiated form, covering the side of the bladder, prostate, and rectum; and are

Inserted into the two last bones of the os coccygis, and into the extremity of the rectum, uniting on the posterior surface of the gut, with the fibres of the opposite side.—The anterior fibres of this muscle are closely united with the fibres of the sphincter ani, passing within them, but on the outside of the longitudinal fibres of the gut itself.

Situation: This muscle, with its fellow, resembles a fleshy funnel, which is pierced below to give passage to the termination of the intestinal canal. Its external surface is separated from the fascia covering the obturator internus by adipose substance. Internally, the levator will be seen, hereafter, lined in part by a dense fascia; its posterior edge is contiguous to the next muscle. From its situation, it assists in closing the lower outlet of the pelvis.

Use: To draw the rectum upwards, assisting in the

expulsion of the fæces, and in closing the gut, after the evacuation : to support the pelvic viscera.

6. The Coccygeus *arises*, by a narrow origin, from the spinous process of the os ischium ; it covers the inside of the anterior sacro-sciatic ligament, expanding to form a thin triangular fleshy belly.

Inserted, by its base, into the extremity of the os sacrum, and into the lateral surface of the os coccygis, immediately before the gluteus maximus.

Situation : It is deep-seated between the levator ani and edge of the gluteus maximus, and cannot be exposed, until the viscera of the pelvis have been removed.

Use : To support and move the os coccygis forwards, and to connect it more firmly with the sacrum. It completes, posteriorly, the fleshy septum at the outlet of the ˴ pelvis, and might be considered as the posterior portion of the levator ani.

The Arteries met with in the dissection of the perineum are, (1.) the *Perineal Artery,* observed in the triangular space between the erector and accelerator muscles, coming forward from the internal pudic, and passing on superficially towards the scrotum, supplying the muscles and integuments of the perineum. 2. The *Transverse Artery of the Perineum,* a smaller artery, often a branch of the perineal, running along the line of the transverse muscle. 3. The *Artery of the Bulb* will be discovered more deeply situated, in the uppermost part of the perineum, crossing it transversely, and entering the bulb of the urethra. 4. The trunk of the *Pudic Artery* itself will be found in a future stage of the dissection, deep-seated, on the inside of the tuber ischii. The filaments of the *Pudic Nerve,* which is sent off from the sacral nerves, and comes out from the pelvis

6

between the two sacro-sciatic ligaments, will be found in company with the arteries in the perineum.

Of *the more* DEEPLY-SEATED PARTS *in the* PERINEUM.

We now proceed to remove the muscles of the perineum, and, by separating the rectum from the posterior surface of the bladder, to display the parts more deeply seated.

Dissect off the acceleratores, or divide them along their middle line of union, and reflect them, as well as the transverse muscles, outwards to their attachments to the bones : this will at once lay bare the *Corpus Spongiosum* of the *Urethra ;* observe its projecting or pendulous part, the *Bulb* of the *Urethra,* which lies immediately beneath the lower fibres of the two accelerators, in the middle line of the Perineum.—On each side of the bulb, will be discovered a firm ligamentous expansion, stretched between the rami of the pubis, of a triangular form ; this is the *Triangular Ligament* of the *Urethra ;* it is seen descending from the symphysis, along the diverging rami of the ossa pubis on each side, filling up the space between the rami, and sending tendinous fibres or processes downwards to the fore-part of the rectum, and also, laterally, to the front edge of each levator ani : this ligament is distinctly felt by the finger, and has a defined crescentic edge below, and the bulb of the urethra is lying on its anterior surface, closely connected with it.

Dissect cautiously below the bulb, and lift it up from the triangular ligament ; you will perceive the *Membranous Part* of the *Urethra* coming through the ligament, the lower part of which it perforates by a circular opening ; and here, immediately under the bulb, search for *Cowper's Glands.*

Prosecute your dissection behind the lower edge of the triangular ligament; separate its fibrous connexion with the anterior part of the rectum, and pull the gut with its sphincter downwards: you first expose the *membranous part* of the *urethra*, extending from the bulb backwards to the apex of the prostate: press your staff downwards, and, at some distance behind the triangular ligament, you discover the *Prostate Gland*, embracing the urethra and neck of the bladder, and connected very closely with the rectum, so that the separation is not easily effected: observe that this gland is invested by a dense capsule-like fascia: this is the *fascia*, or *ligament*, of the *prostate ;* it is quite distinct, and adheres closely to the surface of the gland, and may be considered as a prolongation backwards of the triangular ligament. On each side, immediately behind the triangular ligament, are seen the fibres of the *Levator Ani*, coming down from the inside of the ossa pubis, over the neck of the bladder and side of the prostate, to which parts the inner surface of the muscle is connected by cellular tissue; the anterior borders of the two levators descend parallel to each other, and are separated by the prostate and neck of the bladder, which pass forwards in the slit-like interval between the two muscles :—the muscle on each side is here connected with the perineal muscles; it then passes over the side of the rectum, and unites with its fellow on the back part of the gut, intermixing with the sphincter ani, and having its posterior fibres fixed to the coccyx.—The outer surface of the levator ani is separated, by much cellular and adipose tissue, from the tuber ischii, and from the fascia covering the obturator internus :—dissect the fat away from this deep triangular hollow, and observe the *Obturator Fascia*, descending over its muscle along the inside of the ischium : observe, also, that the outer surface of the levator is likewise covered by a thin fascia, which appears prolonged over it from the triangular ligament. Now search for the *Pudic Artery*, which will be readily discovered, if injected ; you will find it deep-seated, on the inner surface of the tuber ischii, about an inch and a half above the lower flattened part of that process : observe that it is tied down by the fibres of the obturator fascia, and is partly sheltered by the *falciform process*, or floating edge, of the great sacro-sciatic ligament, into which the obturator fascia passes : the artery then ascends along the inside of the rami of the ischium and pubis, and becomes involved in the fibres of the trian-gular ligament.

Continuing to divide the cellular tissue, which connects the rectum with the posterior surface of the blad-der, and retracting the gut downwards, you may bring into view the *Vesiculæ Seminales* at the base of the prostate ; the *Vasa Deferentia*, running between the two vesiculæ ; and the *Ureters* entering the lower part of the bladder. These parts will be thus displayed adhering to the bladder ; but they are more readily examined in the side view of the pelvis, for which purpose the walls of the left side of that cavity may now be removed.

But before taking the lateral view of the Pelvic viscera, it will be expedient to turn to the cavity of the abdomen, and to examine the course of the *Peritoneum* in the pelvis, and the *Fascia Vesicalis*.

- Inflate the bladder, and trace the peritoneum descend-ing from the abdomen : it is seen passing from the recti muscles to the fundus of the bladder, continued over the fundus to the posterior and lateral parts of the viscus, and then passing on to the fore part of the rectum ; it leaves a pouch between these two organs, and forms laterally two *semilunar folds,* which are seen deep in the pelvis, and have been termed the *Posterior Ligaments* of the *Bladder ;* into the pouch so formed, the hand will pass, and it will be found to be from four to five inches above the anus, just beyond the base of the vesiculæ seminales. From this point the peritoneum is reflected upwards, at first covering only the fore part and sides of the rectum ; but it gradually encircles the gut, and, attaching it loosely to the sacrum, is called the *Meso-rectum,* which is continued, above, into the iliac portion of the mesocolon :—from each side of the bladder the peritoneum passes outwards to the walls of the pelvis.

Strip the peritoneum from the left side of the bladder, and from the walls of the pelvis of the same side ; press the bladder, which should be only moderately inflated, in the opposite direction ; and, having removed some loose cellular tissue, look downwards deep into the pelvis ; you will perceive a dense shining fascia, passing from the walls of the pelvis to the neck and lateral surface of the bladder : this is the *Fascia Vesicalis.* Draw back the bladder from the ossa pubis, remove the adipose and cellular tissue interposed ; and you will observe, on each side of the symphysis, a strong fasciculus of shining ligamentous fibres, passing horizontally from the neck of the bladder, to be fixed to the symphysis ; these are named the *Anterior Ligaments* of the *Bladder*,* and the fascia vesicalis appears continued from them on each side, and is also prolonged between the two portions, forming, just under the symphysis pubis, an oval pit-like depression, which will admit the end of the finger. This vesical fascia, however, descending into

---

* Sometimes described singly as the *Inferior Ligament* of the Bladder.

the deepest part of the pelvis, will be traced more readily, when the side of the cavity has been removed. The *Superior Ligament* of the bladder, consisting of the *Urachus,* a ligamentous chord running up from the fundus to the navel, and of two other chords, which are the remains of the *umbilical arteries,* passing up from the sides of the bladder, is seen ascending beneath the stripped off peritoneum.

Of the Side view of the Pelvic Viscera.

Now proceed to remove the bony walls of the pelvis on the left side :—detach the left crus penis from its origin, and saw through the left os pubis half an inch from the symphysis : force apart, by pressing the thighs outwards, the divided edges of the bone, and, as they recede, examine attentively the relative situation of parts : the origin of the left levator ani from the inside of the pelvis is brought into view, and must be divided. Complete the section of the pelvis by cutting through the soft parts, keeping the knife close to the bony walls, and turning the viscera to the right side : you thus expose the left sacro-iliac symphysis ; divide the ligamentous connexion of the sacrum and ilium, and force the two bones asunder ; the left thigh and side of the pelvis are then to be removed.

If the subject is emaciated, very little further dissection will be required; the left levator ani, descending over the side of the bladder and rectum, must be dissected off or reflected downwards ; the bladder well inflated from the ureter; the rectum distended with horse-hair, and any loose fat and cellular tissue removed.

Observe,

1. The Bladder. *Situation :* within the pelvis, immediately behind the ossa pubis, and before the rectum ;

it is covered on the upper and back part by the peritoneum ; in front, and below, it is connected by cellular
tissue to the surrounding parts ; its anterior surface is
applied to the ossa pubis, and there is a considerable
quantity of adipose tissue interposed : its posterior surface lies in the hollow of the curve of the rectum, and,
below the reflection of peritoneum, is connected to the
gut by loose cellular and adipose tissue, in which the
vesiculæ seminales and vasa deferentia lie concealed.—
Shape : oval, but flattened before and behind, and, while
in the pelvis, somewhat triangular, varying also according to the state of distention. **Divided** into the *fundus*
or bottom, *corpus* or body, and *cervix* or neck. In the
contracted state, the fundus is the broadest and roundest
part ; but, when distended, the cervix is broader than
any other part. At the top of the bladder, above the
symphysis pubis, may be remarked its *Superior Ligament*, before described ; and, on the fore part, the two
Anterior Ligaments, fixing it to the symphysis. Observe
the parts of the bladder, not covered by peritoneum, as
they are the situations of surgical operations : these are
the whole *anterior surface*, lying against the pubis, and
rising above it, when the bladder is distended, so that it
may be punctured above the pubis ;—the *sides*, at the
very lowest part of which the cut is made in the lateral
operation of lithotomy, and where the viscus may be
punctured from the perineum ;—and the inferior surface,
resting on the rectum, and allowing us to puncture from
it. Observe also the direction of the axis of the bladder,
in conformity with which all instruments should be introduced : this is a line drawn from the navel to the os
coccygis.

. Examine again the Fascia Vesicalis ; for which purpose the peritoneum may be stripped from the right side

of the bladder, and the fascia of that side traced; it is seen passing from the walls of the pelvis to the side of the bladder, and extending from the symphysis pubis backwards to the sacro-sciátic notch *. Anteriorly, it appears continuous with what has been usually named the anterior ligaments of the bladder; it passes over the edge of the prostate, and from the neck of the bladder along its lateral surface; downwards, it dips deep into the cavity of the pelvis, passing over the side of the rectum, and is then reflected upwards on the walls of the pelvis, where it is evidently continuous with the fascia lining the cavity generally, which is a prolongation of the fascia iliaca. It is united very closely to the side of the bladder, by cellular tissue and a plexus of vessels; and, extending from before backwards, is fixed, behind, to the spine of the ischium, and to the lateral surface of the sacrum, where it becomes thin and indistinct. This fascia lines in part the obturator internus, and also, in part, the inner surface of the levator ani: it divides the deeper from the more external parts of the pelvis.

2. The Prostate Gland. *Situated,* between the bladder and the rectum; surrounding the beginning of the urethra in such a manner, that one third of its thickness is placed above the urethra, and two thirds below it; its *shape* is chesnut-like, and its *consistence* dense and firm. Observe, that its *situation* is, behind the lower part of the symphysis pubis, at the neck of the bladder,

---

* A denser line or band of this fascia is observed, extending from the spine of the ischium to the symphysis pubis, crossing obliquely the surface of the obturator internus. The general fascia of the pelvis appears here to split into two laminæ, of which, one *the vesical,* passes inwards to the side of the bladder, while the external lamina descends over the obturator muscle, forming the *obturator* fascia. The fibres of the levator ani spring, externally, from this angular line of union of the two fasciæ.

which it embraces, as well as the first portion of the urethra; that it lies on the fore part of the rectum, through which it may be felt by the finger, and that it is two inches or more above the anus : observe that its *upper surface* is immediately beneath the anterior ligaments of the bladder, and that it is connected with the symphysis, and with the posterior fibres of the triangular ligament, from which ligament it appears to derive its dense, capsule-like, fascia ; press down the prostate from the symphysis, and you will frequently see some muscular fibres descending, described as the *Compressor Prostatæ* ; the *lower surface* of the prostate is somewhat convex ; its base is placed on the neck of the bladder ; its apex extends along the urethra. It has been divided into two *lateral lobes*, and a third posterior or *middle lobe*, which last is more observable, if the gland be diseased : its *sides* are rounded, and are in contact with the anterior fibres of the levatores ani ; the two vesiculæ seminales are attached to its base ; and its substance is perforated by the *Ejaculatory Ducts*, which are formed, on each side, by the union of the vas deferens with the short duct of the vesicula seminalis : there is a *plexus of vessels*, chiefly veins, surrounding the base of the prostate and neck of the bladder.

3. The URETHRA ; the curve should be carefully observed ; the urethra begins at the neck of the blader, passes forwards under the arch of the pubis, and is then continued to the extremity of the penis, in the groove formed by the two corpora cavernosa. Its divisions are ; (1.) Its beginning, or *Prostatic part*, about an inch in extent, which is embedded in the prostate gland. (2.) The *Membranous part*, which extends from the prostate to the bulb, lying under the symphysis pubis, and perforating the triangular ligament, which fixes this

part of the urethra in its situation. Observe, that this membranous portion of the canal is scarcely an inch in length; it lies above the extremity of the rectum, and is surrounded by dense fibrous tissue, continued from the triangular ligament: there is nearly an inch of space between the urethra and the symphysis pubis, and here the veins of the penis are found passing. Remark also, that, in pulling down this membranous part of the urethra from the symphysis, some fleshy fibres, more or less evident in different subjects, may be observed, descending from the symphysis over the sides of the urethra, so as to enclose it; these are the *compressóres* or *levatores urethræ* of Mr. Wilson; and the compressor muscle of the prostate. is merely a continuation backwards of these fleshy fibres : cut across this membranous portion of the urethra, and you will find its sides formed of a dense, sponge-like, substance. (3.) The urethra, having perforated the triangular ligament, enters the *corpus spongiosum.*

4. The Corpus Spongiosum Urethræ surrounds the canal to its extremity, being situated under and between the united corpora cavernosa; at its beginning, it forms a considerable body of a pyriform shape, termed the *Bulb of the Urethra,* which has been already seen, immediately beneath the lower fibres of the accelerator muscles, adhering to the anterior surface of the triangular ligament.

5. The Corpora Cavernosa Penis arise, on each side, by a process, named the *Crus,* from the rami of the ischium and pubis, as low down as near the tuber ischii; the two crura ascend, and are united, immediately before the lower part of the symphysis pubis; they are covered ·by a strong, white, shining, fibrous membrane, which is very elastic; and they are fixed above

to the symphysis by a dense, triangular, lamella of fibres, named the *Suspensory Ligament*. By the union of the two corpora cavernosa, two grooves are formed : (1.) A smaller one above, in which two arteries pass, a large vein or two betwixt them, and some twigs of nerves. (2.) A larger groove below, which receives the urethra.

6. The GLANDULÆ ANTE-PROSTATÆ, or COWPER'S GLANDS, are two small reddish bodies, of the size of peas, situated at the back part of the bulb, on the sides of the membranous part of the urethra, under the fibres of the triangular ligament : they are often indistinct, but may generally be discovered by the finger.

7. The VESICULÆ SEMINALES, are two soft, whitish, knotted bodies, about two inches and a half in length, and half an inch in breadth ; flattened, and of a conical or pyriform shape. *Situated* between the rectum and lower part of the bladder obliquely, so that their inferior extremities are contiguous, and are affixed to the base of the prostate ; while their superior extremities are at a distance from each other, extending outwards and up-wards, and terminating just on the inside of the insertion of the ureters in the bladder.

8. The two VASA DEFERENTIA are seen running be-twixt the vesiculæ seminales, and united to them, and to the base of the prostate. Observe the *triangular space*, bounded by these tubes and the vesiculæ laterally ; be-hind by the cul-de-sac of peritoneum, and having its apex immediately behind the base of the prostate ; in which the bladder is in contact with the rectum, with only some cellular tissue and vessels, chiefly veins, interposed. Trace the vas deferens in its descent from the internal ring, where it leaves the other vessels of the spermatic chord ; passing in the fold of peritoneum to the side of the bladder, it continues downwards on the posterior and

lower part of that viscus, crossing before the ureter, and approaching the vas deferens of the other side : it is then traced running on the inside of the vesicula seminalis, to the base of the prostate.

9. The entrance of the URETERS, into the lower and back part of the bladder, will be found on the outside of the vesiculæ seminales ; and at the distance of about an inch from the base of the prostate. Trace the ureter descending, behind the peritoneum, over the sacro-iliac symphysis into the pelvis.

10. The RECTUM follows the curve of the sacrum and coccyx, descending behind the posterior surface of the bladder. Continued from the sigmoid flexure of the colon, the rectum commences on the left side of the sacro-vertebral articulation, and is applied, superiorly, to the left half of the anterior surface of the sacrum, but, soon inclining from left to right, it descends along the median line : it is tied to the sacrum by its short peritoneal fold, which passes down over the upper two thirds of the gut ; the lower third of the rectum is united, anteriorly, by cellular tissue, to the bladder, vesiculæ seminales, and prostate ; closely to the prostate, but between the bladder and rectum there is much cellular tissue : posteriorly, the lower part of the rectum is connected loosely with the sacrum and coccyx, and, for some distance, it is covered on its sides and behind by the levator muscles, and, at this point, the gut turns slightly backwards to reach the anus, being separated, in front, from the membranous portion and bulb of the urethra, by a cellular interval. The anus is placed an inch before the coccyx, the extremity of the gut is contracted, and encircled by the sphincter, while, immediately above, it is found dilated, especially in the aged, into a pouch.

Having concluded the examination of the pelvic viscera, you may now cut across the fascia vesicalis of the right side, and, beneath it, you expose the *right levator ani;* clean carefully the internal and external surfaces of the muscle, and its fan-like appearance, passing from the inside of the pelvis to the rectum and coccyx, may be well displayed. Next dissect off the levator, and thus examine the *triangular ligament* from within the pelvis : you will find that this ligament is, laterally, continuous with the obturator fascia ; and also that it does not go up to the symphysis pubis, but that, for half an inch below the symphysis, the space is occupied by the *Pubic Ligament.*

SECTION II.

OF THE VESSELS AND NERVES CONTAINED WITHIN THE PELVIS.

I. ARTERIES.

THE INTERNAL ILIAC or HYPOGASTRIC ARTERY, having left the trunk of the Common Iliac, descends into the cavity of the pelvis, over the sacro-iliac symphysis ; after a short course of two inches, arriving at the upper part of the great sacro-sciatic notch, it terminates by dividing into many branches, which disperse in various directions, supplying the parts within the pelvis, as well as the muscles outside. These arteries are of variable origin, but usually come off from two common trunks, into which the hypogastric divides more or less evidently.

The *Posterior branch,* or subdivision of the hypogastric gives off,

1. The ILEO-LUMBAR ARTERY, a considerable vessel,

which passes outwards beneath the psoas muscle, to ramify chiefly on the iliac fossa. Sometimes this artery comes off in several branches, and its origin also varies.

2. A. Sacræ Laterales, two or three small vessels, which descend on the sacrum, supplying the adjacent parts, and entering the sacral foramina.

3. The Obturator Artery runs forwards immediately under the brim of the pelvis, with the obturator nerve in the same line above it, to reach the upper part of the thyroid ligament, where a notch or foramen, (bounded above by the pubis, below, by the obturator muscle and fascia,) transmits it to the thigh. It gives muscular twigs, within the pelvis, and will be found, in the thigh, covered by the pectineus.

This artery is subject to many anomalies; a frequent variety is, when it arises from the external iliac, by a common trunk with the epigastric; in which case, it turns downwards behind the pubis to reach its foramen, sometimes passing along the margin of the crural ring. The obturator has also been found arising from the gluteal, sciatic, or from the femoral artery in the thigh; but, however distant from its usual situation, it always passes to the obturator foramen.

4. The Gluteal, or Posterior Iliac artery, a very large branch, passes out of the pelvis through the upper part of the sciatic notch, above the pyriformis muscle. It supplies the haunch, and will be found, in the dissection of the thigh, under the gluteus maximus. Before leaving the pelvis, it gives twigs to the obturator, pyriformis, and levator ani muscles.

The *Anterior branch* of the hypogastric gives off the sciatic, pudic, and umbilical arteries, from one of which, or from the trunk itself, proceed the middle hæmorrhoidal, vesical, uterine, and vaginal arteries.

5. The Umbilical Artery, in the fœtus, passes on to the navel, and appears to be the continued trunk of the hypogastric ; but, in the adult, it is only pervious to the side of the bladder, to which it gives branches, and, in the female, also to the uterus. Beyond this point, it is converted into a ligament, which extends upwards to the umbilicus.

6. The Vesical Arteries, variable in their number ; distinguished into the *inferior*, which supply the lower part of the bladder and prostate ; and *superior*, which commonly come from the umbilical.

7. A. Hæmorrhoidalis Media is an uncertain branch, descending in ramifications on the fore part of the rectum, and back part of the bladder in the male, or of the vagina in the female.

8, 9. The Uterine and Vaginal Arteries, which are found in the female. The latter is often wanting, but the uterine artery is constant, passing inwards to the upper part of the vagina, to which and the bladder it gives branches, and then ascends in the broad ligament to the side of the uterus.

10. The Sciatic Artery, a large branch, passes out of the pelvis by the sciatic notch, below the pyriformis muscle. It descends nearly vertically in the pelvis, and turns outwards between the lower border of the pyriformis and the anterior sacro-sciatic ligament : it gives twigs to the muscles within the pelvis, and will be seen in the hip, between the great trochanter and tuber ischii, under the gluteus maximus. The sciatic frequently comes off by a common trunk with the next artery, or with the gluteal.

11. The Internal Pudic Artery is the branch of the internal iliac, which is more immediately destined to 'supply the parts of generation, perineum, and lower part

of the rectum : it goes out of the pelvis above the anterior sacro-sciatic ligament, twists round it, and re-enters
the pelvis, above and before the great or posterior sacro-
sciatic ligament: it then takes its course along the
ischium, above the tuber of that bone, ascends on the
inner surface of the rami of the ischium and pubis, and,
reaching the root of the penis, divides into its terminating branches, being involved in the fibres of the triangular ligament. While in the pelvis, the pudic sends
off branches to the bladder, vesiculæ seminales and rectum, and to the neighbouring muscles ; and also, in the
female, to the uterus and vagina.

In ascending within the tuber ischii to the arch of the
pubis, the pudic gives off,

(*a*,) *A. Hæmorrhoidales Externæ*, one or more small
vessels, passing transversely to the anus and lower
part of the rectum.

(*b*,) *A. Perinei*, or *Superficial Perineal Artery*. This
considerable vessel has been seen in the dissection of
the perineum, coming off a little anterior to the tuber
ischii: it passes forwards, ramifying and becoming
more superficial, and terminates in the septum scroti.

(*c*,) The *Transverse Perineal Artery*, much smaller,
and often a branch of the last, also noticed before.

(*d*,) The *Artery of the Bulb*, a large but short artery,
comes off from the pudic, opposite the bulb of the
urethra, and is enveloped in the fibres of the triangular ligament ; it passes across to the bulb, also giving
a branch to the prostate.

The continued *trunk of the pudic* having now arrived
near the symphysis pubis, divides into its two terminating branches ;

(*e*,) *A. Dorsalis Penis*, the more superficial branch,
passes under the arch of the pubis, and, piercing the

suspensory ligament, runs along the dorsal surface of the penis, parallel to its fellow, to supply the integuments and glans.

(*f*,) *A. Profunda Penis*, or *Artery of the Corpus cavernosum*, passes deep into the crus penis, and ramifies through the corpus cavernosum.

In the female, the Pudic artery sends branches to the parts in the perineum, and to the labia; its two terminating branches go to the clitoris.

II. VEINS.

The HYPOGASTRIC or INTERNAL ILIAC VEIN is seen in the pelvis behind its artery, and is formed by branches, which correspond with those of the artery. The veins, supplying the pelvic organs, ramify largely, and form plexuses, which are named from their situation, as the hæmorrhoidal, pudic, vaginal, &c. The dorsal veins of the penis do not join the pudic vein, but pass under the arch of the pubis, to terminate in a venous plexus on the side of the bladder, named the *Vesical.*

III. NERVES.

In the dissection of the pelvis and back part of the abdominal cavity, we meet with the trunks and branches of the lumbar and sacral nerves.

There are five pairs of LUMBAR, and five or sometimes six pairs of SACRAL NERVES, which emerge from the spinal canal; the lumbar, by the intervertebral foramina; the sacral nerves, by the foramina on the anterior and posterior surfaces of the sacrum. The small *posterior* branches of these nerves pass to the muscles and integuments behind the spine, and to the buttock; but the large *Anterior branches* of the ten nerves unite, and, by their interlacement, form a plexus, termed the

CRURAL PLEXUS, or by some described, as consisting of two parts, the Lumbar or Lumbo-abdominal, and the Sacral or Sciatic plexus. This plexus, in its lumbar part, is situated close to the sides of the lumbar vertebræ, behind the psoas; and it is continued down, by a large communicating trunk, into the pelvis, where the sacral portion lies on the sacrum and pyriform muscle, behind the rectum. It gives off smaller branches, which are formed of filaments of one or two nerves; but the great mass of the crural plexus terminates in the three large nerves of the lower extremity.

The *smaller branches* of the crural plexus are the following:

1. The *Musculo-cutaneous branches*, subject to many variations, but usually three in number, are derived from the two or three first lumbar nerves, passing outwards from under the psoas: (*a*,) The *superior branch*, or *ilio-scrotal nerve*, passes transversely outwards between the peritoneum and transversalis, then pierces that muscle, and runs forwards along the crest of the ilium, giving filaments to the muscles and integuments, and one *long branch*, which follows the direction of the crural arch to the external ring, where it emerges and is distributed to the skin of the scrotum, groin, or labia. (*b*,) The *middle branch* passes outwards in front of the iliacus externus, covered by peritoneum, pierces the transversalis, then the internal oblique; some of its filaments pass to the scrotum and groin. (*c*,) The *inferior branch*, or *Inguino-cutaneous nerve*, is seen passing obliquely over the iliacus, towards the anterior superior spinous process of the ilium; it penetrates the space between that process and the inferior spine, and is then placed beneath the fascia lata of the thigh, where it divides and soon becomes sub-cutaneous.

2. The *Genito-crural Nerve,* or *External Pudic,*
springs from the first and second lumbar nerves, passes
between the upper digitations of the psoas, and is seen,
behind the peritoneum, descending over the anterior
surface of that muscle, then along the outer or fore part
of the external iliac artery; near the crural arch, it
divides into (*a*,) An *inguinal branch,* which turns forwards to accompany the chord through the inguinal
canal, ramifying on the cremaster and its coverings, or
on the round ligament and groin. (*b*,) The *Crural
branch* descends with the iliac vessels under Poupart's
ligament, to ramify in the inguinal region and integuments on the inside of the thigh.

3. The *Superior Gluteal Nerve* comes off from the
lumbo-sacral nerve, (or large communicating branch of
the two last lumbar nerves with the upper sacral;) it
leaves the pelvis, with the gluteal artery, above the pyriformis, and is distributed to the gluteal muscles.

4. The *Inferior Gluteal,* or *lesser Sciatic nerve* is
derived from the sacral nerves, and leaves the pelvis
below the pyriformis; it will be seen, in the back of the
hip, on the side of the great sciatic nerve, dividing
into (*a*,) *Gluteal branches.* (*b*,) A *sciatic branch,* to
the inner part of the thigh and perineum. (*c*,) A long
crural branch, or *Posterior cutaneous.*

5. The *Pudic Nerve* also arises from the sacral nerves,
and passes out of the pelvis below the pyriformis, Joining
the pudic artery between the two sacro-sciatic ligaments,
and having a corresponding distribution; it subdivides
into an *inferior* branch, which gives filaments to the
sphincter and perineal muscles, and a *superior* branch,
which ascends to the symphysis pubis, and becomes the
dorsal nerve of the penis. In the female, the two
branches pass to the labia and clitoris.

6. Small *branches* pass from the sacral plexus to the parts within the pelvis, as the bladder, rectum, vesiculæ and prostate in the male, and to the uterus and vagina in the female : also communicating filaments are sent to the hypogastric plexus, and ganglia of the sympathetic.

The three great *terminating branches* of the crural plexus, which are seen in the pelvis, passing to the thigh, are,

1. The ANTERIOR CRURAL NERVE. This nerve is formed by branches of the first, second, third, and fourth lumbar nerves ; at its origin, it lies under the psoas magnus, and then descends, deeply seated, between the psoas and iliacus ; lower down, it is seen shining beneath the iliac fascia, between the contiguous borders of the two muscles ; then, passing under Poupart's ligament, it emerges, and appears on the outer side of the inguinal artery.

2. The OBTURATOR NERVE is the smallest of the nerves destined to the thigh. It is formed by branches of the second, third, and fourth lumbar nerves ; it emerges from behind the inner edge of the psoas magnus, and descending nearly in a straight line, within the brim of the pelvis, to reach the obturator foramen, it joins its corresponding artery, and passes into the thigh.

3. The GREAT SCIATIC NERVE arises by branches from the fourth and fifth lumbar, and three first sacral nerves, which unite together to form the largest nervous trunk in the body. This great nerve passes between the lower border of the pyriformis and the superior gemellus, and thus escapes from the back part of the pelvis by the sciatic notch. Sometimes one of the branches goes through the pyriformis, and joins the sciatic trunk at the back of the pelvis.

SECTION III.

OF THE SCROTUM, AND OF THE STRUCTURE OF THE VISCERA OF THE PELVIS, AND ORGANS OF GENERATION IN THE MALE.

The scrotum consists externally of a loose, rugose skin ; internally, of a peculiar, dense, cellular tissue, termed the *Dartos,* which forms two distinct sacs, containing the testicles. These sacs, applied to each other along the median line, produce the vertical partition between the two testicles, termed the *Septum Scroti.* The Dartos is continued with the superficial fascia of the abdomen and perineum ; it is fixed laterally to the rami of the ischium and pubis, and, by some, has been considered as a cutaneous muscle.

On dividing the anterior part of the scrotum, on either side of the raphe, we expose,

1. The Testicle, a gland of an oval flattened form, covered by the tunica vaginalis, having, on its upper edge, an appendage, termed *Epididymis.*

2. The Spermatic Chord, connecting the testicle with the abdominal cavity. It consists of, (*a,*) The *Spermatic artery,* entering the upper edge of the testicle. *(b,)* The *Spermatic veins,* which form a plexus above the testicle, then ascend, in four or five branches, around the vas deferens, and unite, within the inguinal canal, into a single vein : this often forms a second plexus, the *Corpus Pampiniforme,* below the kidney. (*c,*) The *Spermatic nerves,* which come from the spermatic plexus of the splanchnic nerve ; and also some filaments from the lumbar nerves. (*d,*) The *vas deferens,* or excretory duct of the testicle ; this is situated in the back

part of the chord, and is distinguished by its firm cartilaginous feel.

These vessels, with some absorbents and the epigastric twig, form the packet of the chord, which is covered, in successive layers: 1. By the integuments; 2. By superficial fascia, prolonged from the abdomen into the dartos; 3. By a fibrous layer, or *fascia propria*, derived from the external oblique tendon, but only to be traced for a short distance; 4. By the *cremaster* muscle, the dispersed fibres of which are only observed in front of the chord, and are lost on the tunica vaginalis and tissue of the scrotum; 5. By the *tunica vaginalis communis*, enveloping the chord, as in a sheath, sending processes between the vessels, and continued, within the inguinal canal, with the funnel-like process of the transversalis fascia, and sub-peritoneal cellular tissue. In the fœtus, the peritoneum is prolonged, within the former tunics, forming a sac or digital process, communicating with the abdomen, enclosing the testicle, but anterior to the spermatic vessels: this, after birth, is obliterated along the chord, a cicatrix-like depression at the internal ring marking the point, at which it was protruded.

STRUCTURE OF THE TESTICLES. Slit open the tunica vaginalis; it forms a loose bag without an opening, in which the testicle is suspended; the inner surface of the bag smooth and shining: this serous bag is the remains or unclosed portion of the peritoneal process or sac. Observe, that it is reflected over the testicle itself, adhering firmly to the subjacent tunica albuginea, and investing the gland, except at its back part, where the vessels enter, so that, (as in other serous cavities,) the testicle is external to the sac. Remark the *Epididymis*, placed on the upper border of the gland, of an oblong shape, and swelled out at its two extremities, termed its *Globus*

Major, and *Minor,* and attached to the testicle by one of its edges. Cut into the substance of the testicle: the *tunica albuginea* is seen to be a strong fibrous, capsule-like, membrane, of an opaque white colour; the substance of the gland pulpy, consisting of numerous minute filiform ducts or tubes, termed the *tubuli seminiferi,* which are separated into bundles, by delicate septa or partitions, that proceed from the inner surface of the tunica albuginea. These tubuli pass to the upper border of the gland, where we observe a narrow, oblong, projecting margin, termed the *Corpus Highmorianum,* or *rete testis.* Here they unite into large trunks, ten or twelve in number, convoluted, and named *vasa efferentia,* which traverse the Corpus Highmorianum, and unite to form the epididymis. Examine the epididymis, which, on removing the tunica vaginalis, will have the appearance of a convoluted tube, and terminates in the *vas deferens,* a single white fibrous vessel, of texture almost cartilaginous, with a canal nearly capillary.

VESICULÆ SEMINALES: the surface is nodulated; when cut into, they appear formed of cells, which contain an opaque, brownish, fluid: they have an *exterior* coat, which is white and dense, and an *internal* mucous coat, which is very thin and slightly rugose; the *duct* is very short, and is joined to the vas deferens at an acute angle, to form the *ejaculatory duct.*

PROSTATE GLAND: its texture is dense and firm, and it gives off numerous small straight ducts, which open into the urethra, on the caput gallinaginis: the canal of the urethra, and the two *ejaculatory ducts,* are found traversing the substance of the prostate.

COWPER's GLANDS, of a reddish colour, and firm consistence, and having each an *excretory duct,* which opens into the urethra.

BLADDER. It has three coats. 1. *External* or *Peri-toneal.* 2. *Muscular*, composed of distinct fibres, which are pale and run in various directions : some of these at the neck of the bladder take a circular direction, and have been described as a *Sphincter vesicæ.* 3. *Inner* or *Mucous* coat, united to the muscular by cellular tissue, and continuous with that lining the ureters and the urethra : to examine this coat, slit open the bladder : observe the inner mucous surface, of a whitish colour, and, when the bladder is empty, having numerous rugæ ; the *openings* of the *two ureters,* oblique and contracted : at the lower part of the bladder, near its neck, a *triangular space* may be observed, extending from the openings of the ureters to the orifice of the urethra, where the mucous membrane is particularly smooth, called by French writers *Trigone Vesical,* bounded by two lateral projections, which Mr. Bell has described as muscles ; and at the anterior angle of. this smooth surface, just behind the urethral orifice, there is a small tubercle or projection of the mucous membrane, to which the term *la luette,* or *uvula vesicæ* has been applied.

URETHRA. The canal is enclosed, successively, by the prostate, bulb, corpus spongiosum, and glans penis : Slit it open from above : Observe 1. Its *mucous membrane,* continued from that of the bladder, of a white colour, reddish near the external orifice ; having numerous small openings, termed *lacunæ,* of which one or two near the external orifice are large :—where the urethra passes through the prostate, remark a *longitudinal eminence,* or *crista,* terminating in a point anteriorly, formed by the mucous membrane, and named *caput gallinaginis,* or *veru montanum ;* in the upper part of which, observe a blind pouch or large lacuna, named *sinus Morgagni,* or *foramen cæcum :* in the middle part of the caput gallinaginis, the openings of the ejaculatory ducts, and, more outwardly, the numerous small orifices of the prostate ducts are made evident, on squeezing the prostate, and vesiculæ seminales ; more forwards, the openings of the ducts of Cowper's glands. 2. Remark that the *diameter* of the canal varies ; at its origin from the bladder, it is large, but immediately contracts ; is dilated in the prostate, contracted in its membranous portion, which is the narrowest part of the canal ; in the bulb it again dilates, and continues large to the base of the glans, where there is an evident dilatation, termed *fossa navicularis;* again generally contracted at the external orifice, where it terminates by a narrow vertical slit.

CORPUS SPONGIOSUM. It surrounds the urethra from the bulb to its extremity ; structure apparently cellular, with a dense, fibrous, investing membrane :—it is expanded at its anterior extremity, to form the *glans penis,* the prominent margin of which is named *corona glandis :* the glans is invested by a fold of integument, mucous within, cutaneous externally, termed the *Prepuce,* which makes a duplicature, from the base of the glans to the orifice of the urethra, named *frænum præputii.*

CORPORA CAVERNOSA begin below by the two crura, which unite and terminate, anteriorly, in a rounded extremity, attached to the base of the glans penis : they are covered externally by their elastic fibrous membrane, which forms a kind of cylindric sheath. Internally they are cavernous, or consist of a reticulated cellular, or *erectile* tissue, in which the vessels of the organ ramify. The two cavernous bodies are separated from each other by a septum, which is perforated, and is not continued as far as the glans, named the *septum pectiniforme.*

CHAPTER III.

DISSECTION OF THE ORGANS OF GENERATION IN THE FEMALE.

PREVIOUS to the dissection, it will be proper to examine the external parts, which are collectively called the *Vulva*, or *Pudendum*.

The MONS VENERIS is a rounded prominence, covered with hair, situated at the lower part of the belly, and arising on each side gradually from the groins; it consists of common integuments, with much cellular and adipose substance, and lies upon the fore part of the ossa pubis. From the inferior part of the mons veneris arise

The LABIA PUDENDI, or *Alæ Majores;* which are continued downwards in the direction of the symphysis pubis, and terminate in the perineum; they are formed exteriorly by the skin, interiorly by the mucous membrane, with cellular tissue interposed: their places of union above and below are termed *commissures:* the lower union has been called *frænulum labiorum,* or *fourchette.*

The *Perineum Anterius* is that portion of the soft parts, which extends from the inferior commissure of the labia to the anus. The *Perineum Posterius* is the space betwixt the anus, and point of the os coccygis.

The longitudinal cavity or fissure, situated betwixt the labia, and extending from the mons veneris to the perineum anterius, is sometimes called the SINUS PUDORIS; it is broader above than below, and contains several other parts.

On separating the labia, we see, immediately below the superior commissure, the CLITORIS; a red projecting body, situated below the arch of the pubis, and partly covered by its *Prepuce*. The prepuce is a loose fold of mucous membrane, continued from the inner surface of the labia, so as to cover the superior and lateral part of the clitoris.—The clitoris resembles the penis of the male, and consists of two cavernous bodies; these, however, cannot be traced in this stage of the dissection. That part of the body, which forms an obtuse projection externally, is called the *Glans*.

The NYMPHÆ, or *Alæ Minores*, are two prominent membranous folds, extending from the glans of the clitoris to the sides of the vagina. Their external side is continued from the inner surface of the labia, and from the prepuce of the clitoris; while their internal surface seems immediately prolonged from the fine, thin, vascular integuments covering the clitoris itself; they are spongy, erectile, and consist internally of cellular and adipose substance.

A little lower, we see the *Orifice* of the URETHRA, or MEATUS URINARIUS; it is situated below the clitoris and arch of the pubis, betwixt the nymphæ, and above the orifice of the vagina; it consists of a small rising prominence, like a pea, which projects at its posterior part, and in the centre of which is a small circular opening. On each side of the orifice of the urethra, we meet generally with the orifices of two mucous glands, which by some are named *Cowper's Glands* of the female.

On separating the lower part of the labia pudendi, we see the VESTIBULUM, a space which leads to the vagina; it is bounded behind by the carunculæ myrtiformes, or by the hymen in virgins; on the sides by the labia;

before, by the perineum anterius, which projects forwards, forming a kind of valve, so that a little pit is formed behind it, which is termed *Fossa Navicularis.*

The HYMEN, or *Circulus Membranosus,* is a thin membranous duplicature, placed just within the orifice of the vagina; it generally has an opening in its upper part: its form is various, frequently semilunar, and then its base is attached to the vestibulum, while its cornua extend upwards as far as the sides of the urethra. After the rupture of the hymen, we see some irregular projections, marking the orifice of the vagina, and termed *Carunculæ Myrtiformes;* they are generally supposed to be the remains of the hymen, but are not exactly in the same situation.

Behind these is the VAGINA, or canal leading to the uterus, at the extremity of which may be felt projecting the Os UTERI, or Os TINCÆ; but it cannot be seen without dissection.

The skin should be now divided on the side of the right labium, and the dissection carried from the groin to the side of the anus: the cellular tissue and fascia must be then removed, to expose the following parts.

We find the CLITORIS consisting of two spongy bodies, termed *Crura,* which unite and form the body. The *Crus* of each side is attached to the inner side of the rami of the ischium and pubis, and is united with its fellow, opposite to the symphysis pubis. The body curves downwards towards the urethra: it is divided internally by the *Septum Pectiniforme,* and is attached to the symphysis by a suspensory ligament; it is invested by a dense, fibrous membrane.

The muscles, which are met with in this dissection, consist of four pairs, and two single muscles.

The ERECTOR CLITORIDIS,
    TRANSVERSUS PERINEI,
    LEVATOR ANI,
    COCCYGEUS,
} on each side.

The SPHINCTER ANI,
    SPHINCTER VAGINÆ,
} two single muscles.

1. The ERECTOR CLITORIDIS is a small muscle, which *arises* from the tuber ischii, and from the inside of the rami of the os ischium and os pubis; passes over the crus of the clitoris, and, becoming tendinous, is lost upon it.

Use: To draw the clitoris downwards and forwards.

Arising from the same point, and surrounded by much cellular membrane, we find,

2. The TRANSVERSUS PERINEI.— Its *origin* is the same as in the male.

It is *inserted* into a ligamentous substance in the perineum anterius, at the point where the sphincter ani and sphincter vaginæ meet. This ligamentous or tendinous substance deserves attention. Here, as in the male, it is the point of union into which muscles are inserted.

Use: To sustain the perineum: this muscle is often wanting.

3. Surrounding the extremity of the vagina, and a small part of the vestibulum, we find the SPHINCTER VAGINÆ; it *arises*, anteriorly, from the crura of the clitoris on each side; encloses the vaginal orifice, and is

Inserted into the ligamentous point of the perineum, where the fibres of each side meet, and are connected with those of the sphincter ani, and transverse muscles.

Use: To contract the orifice of the vagina.

4. The SPHINCTER ANI exactly resembles the same muscle in the male.

5. The LEVATOR ANI has also the same attachments as in the male; but, previously to reaching the rectum, it descends by the side of the vagina, adhering firmly to it, and consequently assisting to constrict and support it.

6. The COCCYGEUS is longer than in the male, from the greater transverse diameter of the inferior aperture of the pelvis.

Under the fibres of the sphincter vaginæ, you will find the PLEXUS RETIFORMIS, a lamella of spongy tissue, with convoluted vessels, chiefly veins, of a dark blue colour, about an inch in breadth, and two or three lines in thickness :—it arises from the sides of the clitoris, passes on each side of the extremity of the vagina, is not continued completely around it, but is lost on its posterior part.

The VAGINA is the canal leading from the vestibulum to the uterus. It is five or six inches in length, slightly curved, interposed between the rectum and inferior surface of the urethra and bladder, and terminating posteriorly by a cul-de-sac or blind extremity, which surrounds the os uteri. The vagina is closely united to the urethra,—to the bladder, more loosely; with the lower part of the rectum it is also connected intimately, but its upper part is separated from the intestine by peritoneum. Laterally, it has much cellular tissue and the levatores ani. The vagina is composed of fibrous substance of some thickness, and it is lined by a mucous membrane, presenting the appearance of transverse *rugæ*, which pass obliquely into two longitudinal ridges or *cristæ*, on the front and back surfaces of the canal.

The UTERUS or WOMB. On slitting up the vagina, we see the *Os Uteri*, a rounded projection, with a transverse fissure; the lips of the uterine orifice are distinguished into anterior and posterior, of which the latter is more projecting. The uterus may be examined from the abdomen. It is situated in the middle of the pelvic cavity, between the bladder and rectum, to both of which it is connected by the reflected peritoneum ; it is of the shape of a pear, somewhat flattened, inwardly hollow, outwardly of a whitish colour, and of a firm consistence. The broad upper part is called the *Fundus Uteri*, the narrower part is named the neck or *Cervix Uteri*, and the intermediate part the *Body*. It will be seen that the *Peritoneum* is continued, from the back part of the bladder, to the front of the uterus, that it covers the whole anterior and posterior surface of the uterus, and also a portion of the posterior surface of the vagina, from which it passes on to the rectum, forming a pouch between, as in the male, with two lateral folds.

The uterus has four ligaments, two on each side :

1. The LIGAMENTUM TERES, or *Round Ligament*, is a long, and rounded, fibrous chord, prolonged from the lateral part of the fundus uteri, between the laminæ of the broad ligament, and forming a half circle, to reach the inner opening of the inguinal canal, through which it passes, (as the spermatic chord in the male,) to be lost in the mons veneris, groin, and labia. 2. The LIGAMENTUM LATUM, or *Broad Ligament*, is a broad fold of peritoneum, reflected from the body of the uterus, and connecting it with the sides of the pelvis. The uterus, together with its two broad ligaments, forms a transverse septum, dividing the pelvis into an anterior and posterior half ; in the former of which is the bladder, and in the latter the rectum. The duplicature of the broad ligament encloses the *Fallopian tube, ovary*, and *round ligament*, with the vessels and nerves belonging to those parts.

The FALLOPIAN TUBES are two. Each tube is contained in the upper part of the doubling of the broad ligament; it goes out from the upper angle of the uterus, and is a slender conical tube, four or five inches in length, communicating with the abdominal cavity by its outer end, which is curved downwards and backwards, and terminates by a broad fringed extremity, termed *Morsus Diaboli*, or *the Fimbriæ.*—This broad extremity is connected, by one of the longer fimbriæ or processes, to the next organ.

The OVARIA are two small oval bodies, white and flattened, situated by the sides of the uterus, and enclosed in the posterior fold of the broad ligament, behind the Fallopian tube; each ovarium is connected to the fundus uteri, by a short round fibrous chord, the *ligament* of the *ovary.*

The BLADDER is placed before the uterus, its situation corresponding generally with that of the male, as described in the preceding chapter. It usually rises higher above the pubis, and is of a more rounded figure than in the male.

The URETHRA is short in females, being somewhat more than an inch in length, and capable of great dilatation; its posterior surface adheres strongly to the walls of the vagina and may be felt as a prominent line; it is slightly curved, with the convexity upwards; near the bladder, it is surrounded by a spongy, fleshy, substance. The urethra is connected to the symphysis pubis by the *triangular ligament,* and also to the corpora cavernosa of the clitoris; and some *fleshy fibres* may be observed, as in the male, proceeding from the under part of the symphysis pubis, to spread over the side of the urethra, and encircle it.

The URETER descends from the kidney, over the psoas muscle: it runs for some space betwixt the bladder and vagina, and at last perforates the bladder near the neck.

The RECTUM lies behind the uterus, and vagina: above, the gut is separated from these organs by the fold of peritoneum, which forms a deep cul-de-sac; but, below, the vagina and rectum are united by close cellular tissue and a vascular net-work, forming the *recto-vaginal septum*.

In this dissection, the *Vesical Fascia* is seen, as in the male, passing from the walls of the pelvis to the neck of the bladder, and descending so as to be continued also on the vagina. The *Pudic artery* is met with, having the same general distribution as in the male; the superficial artery of the perineum passes to the labium, and is of considerable size, while the Pudic trunk ascending to the clitoris is small. In the same way, the inferior branch of the Pudic nerve, in the female, is large, but the superior or nerve of the clitoris, of smaller size. The left side of the pelvis may be removed, as in the dissection of the male subject, and the parts examined in the lateral view.

STRUCTURE OF THE UTERUS. Make an incision, and expose the cavity of the uterus; it is triangular, corresponding with the external form of the organ, but very small: externally, the uterus is enclosed between the peritoneal folds; internally, it is lined by a very fine mucous membrane, red and nearly smooth. Between these two membranes, observe the peculiar tissue or *parenchyma* of the uterus, of considerable thickness, firm, elastic, fibrous, and perforated by a great number of vessels: remark the *orifices* of the *Fallopian tubes*: slit open the *cervix* uteri; it has two small longitudinal ridges on its inner surface, with some transverse rugæ.

OVARIA. Observe their surface, irregular and wrinkled, and often exhibiting the appearance of *cicatrices*. Make a section; you find the ovarium covered externally by peritoneum, and under this, but closely adherent, a dense whitish fibrous membrane, which forms a kind of capsule, and sends off cellular processes internally. The *parenchyma* or substance is of a reddish brown, abundantly supplied with vessels, but having considerable firmness; it contains from fifteen to twenty *vesicles*, called the *vesicles of De Graaf*, which are filled with a serous fluid, and sometimes are seen projecting on the surface of the ovarium: occasionally one or more small yellowish bodies, of a glandular appearance, are observed, named the *corpora lutea*.

CHAPTER IV.

DISSECTION OF THE THIGH.

SECTION I.

OF THE ANTERIOR PART OF THE THIGH.

§ 1. OF THE FASCIA, CUTANEOUS VESSELS, AND NERVES.

BENEATH the integuments common to every part of the body, you will find a strong fascia, or aponeurotic expansion, investing the whole thigh. This expansion is named the FASCIA LATA FEMORIS; it surrounds the limb, covering all the muscles, and is pierced by many small foramina for vessels and nerves. It is very strong, smooth, and tendinous on the outer part of the thigh; but, on the anterior and inner part, it is very thin, and of a cellular texture. Therefore, if you wish to demonstrate the whole extent of this fascia, it should be first exposed on the outside of the thigh, and the dissection should be continued very carefully inwards on the fore part, where it is not readily distinguished from the cellular membrane.

7

But, in removing the common integuments from the fore part of the thigh, it will be proper to attend to some parts, which are situated above the fascia.

1. The fascia of the thigh, at its upper part, is covered by a more superficial expansion, which lies over the lower part of the external oblique muscle, completely covers the crural arch, and descends some way beyond the bend of the thigh. This has been called the SUPERFICIAL FASCIA, and has already been described in the dissection of the parts in the groin: in some subjects, it may be traced to the lower part of the thigh.

2. The VENA SAPHENA MAJOR or INTERNA is seen running up on the inside of the knee and thigh. At first it lies very superficial, but, as it ascends the thigh, it is gradually enveloped by the fibres of the superficial fascia, continuing, however, exterior to the fascia lata itself: at the upper part of the thigh, about an inch and a half below Poupart's ligament, it sinks beneath the falciform process of the fascia lata, to join the femoral vein, as described in the dissection of the groin. In its course, it is joined by several cutaneous veins from the thigh, and also from the abdominal parietes, and external organs of generation.

3. Immediately under the true skin, and more superficial than the veins or nerves, you may occasionally perceive the LYMPHATIC VESSELS, running, like lines of a whitish colour, to enter the inguinal glands. These glands are placed in two layers:—the superficial, six or eight in number, are subcutaneous or interlaced with the superficial fascia, about Poupart's ligament and the saphenic opening of the fascia lata, while three or four deeper glands are found close on the inguinal artery, and in the net-work of the cribriform lamella.

4. Several CUTANEOUS NERVES are seen ramifying

above the fascia. They either come from the lumbar. nerves, or from the anterior crural; of the former class, the *Inguino-cutaneous* nerve is found in the external angle of the groin, coming out between the two iliac spines, then dividing into two branches, which pierce the fascia lata, and are distributed to the skin on the outer part of the thigh, while the *genito-crural* branch, entering the thigh with the great vessels, ramifies in the inguinal region and integuments on the fore part. The *cutaneous branches* of the *anterior crural* are three or four in number, two of which usually pierce the fascia just below Poupart's ligament, and descend on the anterior and inner part of the thigh.

The fascia lata may now be fully exposed. Observe, how extensively it arises from the bones, tendons, and ligaments. On the anterior and superior part of the thigh, it arises from the fore part of the crista ilii, from Poupart's ligament, and from the os pubis: on the inside of the limb, it springs from the descending ramus of the os pubis, and from the ascending ramus and tuberosity of the ischium;—behind, and on the outside, from the surface of the sacrum and coccyx, being also continuous with the condensed cellular tissue covering the fibres of the gluteus maximus, which muscle has not, however, any distinct aponeurotic covering. It receives a number of fibres from a muscle belonging to it, viz. the tensor vaginæ femoris, and also from the tendon of the gluteus maximus. From this extensive origin, the fascia lata passes down over the whole thigh, covering and enclosing the muscular mass, and sending septa or processes inwards, which form cellular sheaths for several of the muscles; but the great mass of fascia is unadherent, and does not give origin to muscular fibres. It is firmly fixed to the outer line of the linea aspera by a process or lamina, which passes off from its inner surface: below, it adheres to the common tendon of the rectus and vasti muscles, and to the aponeurotic expansion covering the knee-joint, and it is continued over the knee, to be attached to the heads of the tibia and fibula, after which it forms the fascia of the leg.

The *Semilunar edge*, or *Falciform process*, of the fascia lata at the upper part of the thigh, with the *oval interval* in the fascia at this part, has been described in the dissection of the groin ; it is here that the vena saphena joins the femoral vein.

The fascia should now be dissected back : and in lifting up the thicker part of it, which covers the outside of the thigh, you may observe that it is composed of two laminæ of fibres : the fibres of the outer lamina run in circles round the thigh, while those on the inside, which are stronger, and more firmly connected, run longitudinally. Observe that, at the superior and outer part of the limb, a *strong process* is sent off from the fascia lata, which passes beneath the tensor vaginæ muscle, to be connected with the capsular ligament of the hip-joint, and with the outer tendon of the rectus femoris, ascending also upwards behind the tensor muscle : and, at the lower and external part of the limb, you trace the *strong lamina*, passing from the inner surface of the fascia lata, to be implanted into the outer edge of the linea aspera ; this lamina descends between the vastus externus, and short head of the biceps, being closely connected with both muscles.

§ 2. MUSCLES SITUATED ON THE FORE PART AND INSIDE OF THE THIGH.

These are nine in number.

1. The TENSOR VAGINÆ FEMORIS—*Arises*, by a nar-

row, tendinous and fleshy origin, from the external part of the anterior superior spinous process of the os ilium : it forms a considerable fleshy belly, which becomes broader and thinner, as it descends.

Inserted into the inner side of the great fascia, where it covers the outside of the thigh, and a little below the trochanter major.

Situation : Its origin lies between the origin of the sartorius, and the anterior fibres of the gluteus medius, betwixt which muscles it descends ; it does not lie exterior to the fascia lata, but is completely enclosed in the duplicature, which is formed by the external layer of the fascia, and by the strong process sent off from its inner surface to the capsule of the hip. The fascia of the thigh, where it is continued from the union of these two layers, and from the fibres of the tensor muscle, is very strong, and may be traced, along the outer part of the thigh, in the form of a broad band, to the knee.

Use : To render tense the great fascia of the thigh, to assist in the abduction of the thigh, and in its rotation inwards.

2. The SARTORIUS—*Arises,* by short tendinous fibres, from the anterior superior spinous process of the os ilium ; soon becomes fleshy ; extends obliquely across the thigh, and passes behind the inner condyle : then turns obliquely forwards, and is

Inserted, by a broad and thin tendon, into the inner side of the tibia, immediately below its anterior tubercle.

Situation : Its origin lies between that of the tensor vaginæ femoris, and the outer attachment of Poupart's ligament, and above the anterior fibres of the iliacus internus. It is the longest muscle in the body, and lies before the muscles of the thigh, crossing them like a strap about two inches in breadth ; it runs down for some space upon the rectus femoris, passes over the vastus internus, and then over the triceps adductor longus. At the lower part of the thigh, it runs between the tendon of the triceps adductor magnus, and that of the gracilis. The sartorius covers the femoral vessels in the middle third of the thigh, and, in all this course, is bound down by the fascia lata, which gives to the muscle a complete cellular sheath. It is inserted above the tendons of the gracilis and semitendinosus, over which it sends a broad aponeurotic expansion, to the anterior border of the tibia; and it is also intermixed with that portion of the fascia lata, which invests the knee, and with the fascia of the leg.

Use: To bend the leg obliquely inwards on the thigh, and the thigh forwards, so as to cross the limbs.

3. The RECTUS FEMORIS—*Arises*, by a strong straight tendon, from the inferior anterior spinous process of the os ilium; and, by another strong tendon, which is broader and incurvated, from the dorsum of that bone a little above the acetabulum, adhering to the capsular ligament of the hip-joint. The two tendons soon unite into one, and send off a fleshy belly, which, gradually increasing in breadth and in thickness, runs down over the middle of the anterior part of the thigh. It forms a strong penniform muscle, and again contracting itself at the lower part of the thigh, terminates in a flat, but strong tendon, which is united on each side to the vasti, and is

Inserted into the upper border of the patella; where a thin aponeurosis is sent from it over the fore part of that bone, to be attached to the ligamentum patellæ. This aponeurosis, in union with similar expansions from the two vasti, covers the capsule of the knee-joint at its fore part and sides, and terminates in the fascia of the leg: there is a bursa mucosa interposed between the skin, and that portion of the aponeurosis covering the patella.

Situation : This muscle lies, above, on the capsular ligament of the hip-joint; to expose its tendinous origins, the sartorious and tensor vaginæ femoris must be raised ; and then that tendon, which proceeds from the inferior spinous process, may be seen; partly covered by the outer edge of the iliacus internus ; while the other tendon is exposed, by raising the anterior and inferior fibres of the gluteus minimus. The fleshy belly, at its uppermost part, is covered by the sartorius, and, to allow that muscle to slide over it, is tendinous : below this, it is situated superficially, immediately under the fascia lata, running down over the vasti and cruræus ; and on its posterior surface, where it is in contact with those muscles, it is tendinous. Its insertion lies betwixt the two vasti.

Use : To extend the leg on the thigh, and to bend the thigh on the pelvis : to bring the pelvis and thigh forwards to the leg.

Under the rectus, and partly concealed by it, there is a large mass of flesh, covering the front and sides of the femur, which, at first sight, appears to form but one muscle. It may, however, be divided into three ; the separation on the external surface is not generally very evident, but, by following the course of the vessels, which enter this mass, and by cutting through perhaps a few fibres externally, you will discover the line of separation; and this separation, as you proceed deeper with your dissection, will become very distinct. The three muscles are named, vastus externus, vastus internus, and cruræus : at the upper and middle part of the thigh, they may be separated very distinctly; but, for two or three inches above the condyles, they are connected inseparably.

4. The VASTUS EXTERNUS—*Arises*, tendinous and fleshy, from the anterior surface of the root of the trochanter major, from the outer edge of the linea aspera, its whole length,—from the oblique line running to the external condyle,—and from the whole external flat surface of the thigh bone. The fleshy fibres run obliquely forwards, and downwards: the superior fibres are very long: the inferior, shorter and more transverse.

Inserted into the outer lateral surface of the tendon of the rectus cruris, and into the side of the patella :— Part of it ends likewise in an aponeurosis, which passes over the outer side of the knee to the leg, and is firmly fixed to the head of the tibia, adhering by cellular tissue to the capsule of the knee-joint.

Situation: This muscle forms the large mass of flesh on the outside of the thigh; it is in part concealed by the rectus : on its outer surface, it appears tendinous at its upper part, and fleshy lower down; on its internal surface it is fleshy above, and tendinous below ;—it laps over the outside of the cruræus; where it arises from the linea aspera, it is situated anterior to the tendinous insertion of the gluteus maximus, and to the origin of the short head of the biceps flexor cruris.

Use: To extend the leg, or to bring the thigh forwards upon the leg.

5. The VASTUS INTERNUS—*Arises*, tendinous and fleshy, from the fore part of the root of the trochanter minor, from all the internal edge of the linea aspera, from the oblique line running to the inner condyle, and from the inner surface of the thigh bone. Its fibres descend obliquely downwards and forwards.

Inserted into the inner lateral surface of the tendon

of the rectus cruris, and into the side of the patella; it also sends off an aponeurosis, which is continued down to the leg, and covers the inner part of the capsule of the knee.

Situation: This muscle embraces the inside of the femur, in the same manner as the last-described muscle does the outside, but it is much smaller: it is also in part covered by the rectus. At its upper part the sartorius passes over it obliquely; it laps over the cruræus, and is separated from it with greater difficulty than the vastus externus is. Where it arises from the root of the trochanter, it lies anterior to the common tendon of thé iliacus internus and psoas magnus; and where it arises from the linea aspera and oblique line, it is situated anterior to, and in contact with the insertions of the pectineus and triceps adductor femoris; at the middle and lower part of the thigh, it is closely adherent to the tendons of the adductor longus and magnus, assisting to form the tendinous canal for the femoral vessels.

Use: The same as the last.

6. The Cruræus, or Cruralis—*Arises*, fleshy, from between the two trochanters of the os femoris, from all the fore part of the bone, and from the outside, as far back as the linea aspera; but from the inside of the bone it does not arise, for, on the fore part and inside of the femur, there is a smooth plain surface, of the breadth of an inch, extending nearly the whole length of the bone, from which no muscular fibres arise.

Inserted into the posterior surface of the tendon of the rectus, and upper edge of the patella.

Situation: The principal part of this muscle is lapped over, and concealed, by the bellies of the two vasti; and the small part, which is seen projecting between the anterior edges of those muscles, lies behind the belly of the rectus cruris : between the common tendon of the rectus and cruræus, the fore part of the femur, and the capsule of the knee-joint, there is a large bursa mucosa, which, in many subjects, opens into the joint. A small cellular cavity is also found, between the lower tendinous fibres of the cruræus and the rectus tendon, just before their intimate union. The vasti and cruræus are described by some as a single *Triceps :* or, with the rectus, as a *Quadriceps.*

Use : The same as the last.

A few muscular fibres found under the lower part of the cruræus, and attached to the upper and fore part of the synovial capsule of the knee, have been described as a distinct muscle, the SUB-CRURÆUS. These fasciculi raise the synovial membrane in extension of the leg.

7. The GRACILIS—*Arises,* by a broad thin tendon, from the lower half of that part of the os pubis, which forms the symphysis, and from the inner edge of the descending ramus. It soon grows fleshy, and forms a flat belly, which proceeds downwards on the inner side of the thigh, becoming narrower as it descends. The fleshy fibres pass successively into a rounded tendon, which descends immediately behind the sartorius, turns round the inner condyle of the thigh-bone, and is reflected forwards, to be

Inserted into the inner side of the tibia, under the sartorius.

Situation : It arises from the os pubis on the inside of the origins of the triceps adductor femoris, lying between them and the crus penis : from the pubis to the knee, it runs immediately under the integuments on the inside of the thigh, with its flat surface applied to the adductors, and one edge directed forwards, the other backwards : it is inserted below the tendon of the sartorius, and above that of the semitendinosus ; it is united with the latter tendon, and passes under the expanded termination of the sartorius : the gracilis also gives off some fibres to the fascia of the leg.

Use : To bring the thigh inwards and forwards, and to assist in bending the leg.

8. The PECTINALIS, or PECTINEUS—*Arises,* fleshy, from that ridge of the os pubis, which forms the brim of the pelvis, and from the concave surface below the ridge : it forms a thick flat belly.

Inserted, by a flat tendon, into the linea aspera, immediately below the lesser trochanter.

Situation : Its origin lies on the inside of the belly of the psoas magnus, where that muscle slides over the brim of the pelvis, and on the outside of the origin of the adductor longus :—It descends between the lower edge of the psoas, and the upper edge of the adductor longus : and it is inserted between these two muscles, and posterior to the origin of the vastus internus. Its anterior surface is covered by the fascia lata, and by the femoral vessels : its posterior surface is applied to the obturator vessels and nerve, external obturator muscle, and capsular ligament of the hip-joint.

Use : To move the thigh forwards and inwards, and to perform rotation, by turning the toes outwards.

The insertion of the common tendon of the iliacus internus and psoas into the trochanter minor, is now exposed, and there is a large bursa mucosa between this united tendon, and the capsule of the hip-joint.

9. The TRICEPS ADDUCTOR FEMORIS consists of three distinct muscles, which, passing from the pelvis to the thigh, lie in different layers upon one another, and have nearly the same action.

(1.) The ADDUCTOR LONGUS—*Arises,* by a short strong tendon, from the upper and inner part of the os pubis, near its symphysis; it forms a large triangular belly, which, as it descends, becomes broader, but less thick.

Inserted, tendinous, into the middle part of the linea aspera, occupying rather more than one third of its length.

Situation: It is placed before the two other adductors, at the inner and upper part of the thigh. It arises between the pectineus and gracilis, and above the adductor brevis, being covered by the fascia lata:—The upper edge of its belly ranges with the lower edge of the pectineus; and its insertion lies posterior to the origin of the vastus internus, and anterior to the insertion of the adductor magnus, with both of which muscles it is strongly united, assisting to form the canal for the femoral vessels.

(2.) The Adductor Brevis—*Arises*, fleshy and tendinous, from the os pubis, along the space from the lower part of the symphysis to the foramen thyroideum;—it forms a fleshy belly, becoming broader and thinner.

Inserted, tendinous, into the upper third of the linea aspera, from the trochanter minor downwards.

Situation: It is placed behind the preceding muscle and is smaller : its origin lies under the origins of the pectineus and adductor longus, and on the outside of the tendon of the gracilis :—Its belly descends behind the belly of the pectineus, and behind the superior fibres of the adductor longus ; and its largest part is therefore concealed, but a small portion appears between the lower edge of the pectineus, and upper edge of the adductor longus :—It is inserted behind these two muscles, but before the adductor magnus, and it is much intermixed with all the three muscles at its insertion.

(3.) The ADDUCTOR MAGNUS—*Arises,* principally fleshy, from the lower part of the body and from the descending ramus of the os pubis, and from the ascending ramus of the ischium, as far as the tuberosity of that bone. The fibres run outwards and downwards, having various degrees of obliquity.

Inserted, fleshy above, and tendinous below, into the whole length of the linea alba, and into the oblique ridge above the internal condyle of the os femoris; and, by a roundish long tendon, into the upper part of that condyle. The insertion of this muscle is closely united with those of the two other adductors, and with the vastus internus.

Situation : The superior fibres are nearly transverse, running along the lower edge of the quadratus femoris to the upper fourth part of the linea aspera, and frequently giving the appearance of a distinct muscle. The middle more oblique fibres have a tendinous insertion into the lower three fourths of the linea aspera, at the termination of which line, this tendinous insertion appears to bifurcate, forming an oval opening, through which the femoral artery and vein enter the ham : the inner fibres of the muscle are very long, and descend vertically, ending in the long tendon, which is much intermixed with the vastus internus. This large muscle arises behind and below the two other adductors : it forms a flat partition betwixt the muscles on the fore and back parts of the thigh ; its insertion lies behind the insertions of the long and short adductors, and on the inside of the tendinous insertion of the gluteus maximus, and of the origin of the short head of the biceps flexor cruris.

Use : To draw the thigh inwards, and to approximate the two thighs to each other ;—to roll them outwards.

The two anterior heads will bend the thigh ; the posterior will extend it, when bent forwards.   These muscles
also stady and support the pelvis on the thigh.

§ 3. OF THE VESSELS AND NERVES ON THE FORE PART AND INSIDE OF THE THIGH.

I. ARTERIES.

THE FEMORAL ARTERY may be said to take its course
along the inside of the thigh ;—where it emerges from
under Poupart's ligament, it lies cushioned on the fibres
of the psoas magnus, is called the *Inguinal* artery, and
is very nearly in the mid space between the angle of the
pubis, and the anterior superior spine of the ilium,
nearer however by a finger's breadth to the former.
Having left the groin, it assumes the name of *Femoral*,
and in its course down the thigh, runs over the following
muscles ;—the pectineus, part of the adductor brevis,
where that muscle projects betwixt the pectineus and
adductor longus ; the whole of the adductor longus, and
about an inch of the adductor magnus.   In this course,
after quitting the edge of the psoas muscle, it has the belly
of the vastus internus on its outer side, lying between
the origin of that muscle, and the tendinous insertions
of the muscles over which it crosses, and there is a strong
interlacing of the tendinous fibres ; the artery is invested,
above, by its firm cellular sheath, but, about the middle
of the thigh, it gradually sinks under a broad and distinct fascia, which stretches across from the outer tendinous surface of the vastus internus to the tendons of
the adductor magnus and longus : an *oblique tendinous
canal* is here formed, anteriorly by the broad fascia just

noticed, posteriorly by the interwoven tendons of the vastus and adductors : this canal is nearly three inches in length, and it conveys the femoral artery and vein into the ham, which cavity they enter by passing through the oval interval formed in the tendon of the adductor magnus. To find the artery in any part of its course from the crural arch to the tendinous opening, turn out the foot, ascertain the place of the vessel at the crural arch, and then draw a line downwards to the inner edge of the patella. For some inches below Poupart's ligament, the artery is on its fore part only covered by integument, superficial fascia, some lymphatic glands, and the general fascia of the thigh ; but, meeting with the inclined line of the sartorius, it is, during the rest of its course, covered by that muscle. It perforates the tendon of the adductor magnus, at the distance of rather more than one-third of the length of the bone from its lower extremity.

In its course down the thigh, the Femoral artery is accompanied by the *Femoral Vein.* At the groin, the artery has its great vein on its inner or pubic side ; but, as it descends, the vein inclines gradually backwards, and, at the middle of the thigh, it is found behind the artery, adhering closely to it, and passing with it through the tendinous canal into the ham. Two *filaments* of the anterior crural nerve are generally found accompanying the artery in the middle of its passage down the thigh ; of these one runs before the artery, and then on its inner side, and passes on the inner edge of the sartorius, to be lost in the vastus internus about the knee : the other, a more remarkable branch, the *Nervus Saphenus Internus,* enters the sheath of the artery, lying on the outer and anterior part of the vessel ; and accompanies it in the tendinous canal, but does not pass with it into the ham.

BRANCHES OF THE FEMORAL ARTERY *.

1. Two or three small vessels pass off, just below Poupart's ligament, to the external organs of generation, the *external Pudic*; some twigs are also given to the inguinal glands, skin, and adjacent muscles.

2. The *External* or *Superficial Epigastric* is a slender artery, but constantly found; it comes off half an inch below Poupart's ligament, ascends on the fore part of the expanded tendon of the external oblique, under the superficial fascia, and is traced as high as the navel.

3. The *Arteria Profunda*, or **Deep** *artery* of the thigh, comes off from the femoral artery, at the distance of two to three inches below Poupart's ligament; it is nearly as large as the femoral artery itself, and lies concealed behind it, taking its course downwards and backwards, between the triceps adductor and vastus internus muscles, and terminating, about the middle of the thigh, in three or four branches, which perforate the adductor muscles to the back part of the limb. The branches of the Profunda are,

(*a,*) The *Internal Circumflex artery*; this branch encircles the upper part of the thigh on the inside, passing deep between the pectineus and tendon of the. psoas, round the inner side of the neck of the os femoris; after giving muscular twigs, it divides behind the neck of the bone, into an *ascending* branch, which gives an artery to the cotyloid cavity, and a transverse or *inferior* branch, which ramifies on the muscles behind the great trochanter, anastomosing with the obturator, gluteal, and other arteries of the hip and thigh.

---

* The Epigastric and Circumflexa ilii were before described as branches of the inguinal artery, or rather of the external iliac.

(b,) The *External Circumflex* takes its course transversely outwards, beneath the sartorius and rectus
muscles, and soon subdivides into two branches, of
which one, the *descending*, passes down on the posterior surface of the rectus, and is distributed to that
and the adjoining muscles ; while the other *transverse*
branch bends round the upper part of the femur, to
the muscles on the outer and back part of that bone,
inosculating with the internal circumflex and other
arteries.

(c,). The *Perforating Arteries* are three or four large
branches, in which the Profunda terminates ; these
pass backwards, perforating the triceps adductor, and,
ramifying in various directions, supply the great mass
of muscles situated in the back part of the thigh,
where they inosculate with one another, and with the
circumflex and arteries of the hip.

4. *Muscular* branches come off from the femoral in
its course down the thigh, but are inconsiderable in size.

5. *Ramus Anastomoticus Magnus*, is sent off just before the artery passes through the tendon of the triceps,
and takes its course downwards on the vastus internus,
penetrating its substance, and inosculating with the arteries about the knee.

This is the usual distribution of the arteries of the
thigh, but there are occasional varieties ; the profunda
is constant, though it sometimes comes off higher or
lower than usual, but the great muscular branches are
sometimes given off from the trunk of the femoral ; they
all unite largely with one another, and with the descending branches of the internal iliac, forming a chain of
anastomoses.

The OBTURATOR ARTERY, which is a branch of the
internal iliac artery, passes through the notch at the

upper part of the forámen thyroideum, accompanied by its vein and the obturator nerve; it comes into the thigh, along the upper border of the external obturator muscle, behind the pectineus, and immediately divides into two branches, of which, one, *anterior*, ramifies among the adductor muscles; and the other, *posterior*, passes downwards between the two obturator muscles, to terminate about the tuber ischii.

II. VEINS.

The course of the FEMORAL VEIN in the thigh has been described; its branches correspond with those of the femoral artery, but about an inch below Poupart's ligament, it receives the *Vena Saphena Major*, to which there is no corresponding artery, then entering the abdomen, it becomes the external iliac vein.

The OBTURATOR VEIN accompanies the obturator artery, and has the same distribution.

III. NERVES.

The ANTERIOR CRURAL NERVE, where it passes from under Poupart's ligament, lies about half an inch on the outside of the femoral artery; it immediately divides into a great number of branches, which supply the muscles and integuments on the fore part and outside of the thigh. Some of these branches are superficial, and, after descending for some distance under the fascia lata, pierce it to be distributed to the integuments: other larger branches pass into the neighbouring muscles on the upper part and middle of the thigh; the two filaments of the anterior crural nerve, which descend with the femoral artery, under the sartorius, along the thigh, have been already noticed; one of these, the *Nervus Saphenus Internus*, or *Cutaneus Longus*, is a considerable nerve; where the femoral artery is about to pass into the ham, this nerve emerges from the tendinous canal, runs behind the sartorius, and appears as a cutaneous nerve on the inside of the knee, between the tendons of the gracilis and sartorius; then proceeding downwards on the inside of the leg, it is largely distributed over the tibia, is intricated with the vena saphena major, and terminates on the inner ankle and upper part of the foot.

The OBTURATOR NERVE is found accompanying the obturator artery and vein, lying deep between the pectineus and obturator externus muscles; it divides into two branches, of which (*a*,) the *anterior* and larger, is distributed to the gracilis, long and short adductors, and sometimes gives a large communicating branch to the N. Saphenus Internus. (*b*.) The *Posterior* and smaller branch gives filaments to the obturator muscle, and descends between the adductor brevis and magnus, terminating in the lower part of this last muscle.

SECTION II.

OF THE POSTERIOR PART OF THE HIP AND THIGH.

§ 1. OF THE FASCIA.

THE great mass of muscular fibres forming the buttock is not covered by any distinct fascia; it consists of one thick muscle, named the Gluteus maximus, the fibres of which are superficial, and are exposed on removing the integuments and adipose substance. The fascia lata commences at the anterior edge, and on the lower fibres of this muscle, and descends investing the whole posterior part of the thigh, to be continued on the leg.

Several *cutaneous nerves* are met with in this dissection. In the hip, these nerves are derived from the musculo-cutaneous branches of the lumbar nerves, coming over the spine of the ilium, or from the posterior branches of the sacral nerves, through the sacral foramina. In the back part of the thigh, two or three considerable nerves are observed, piercing the fascia, below the fold of the gluteus maximus, and descending in long branches to the ham : these *posterior cutaneous nerves* are furnished by the great sciatic nerve, and by the gluteal or lesser sciatic nerve, which latter nerve has usually a long crural branch, descending for some distance beneath the fascia, then piercing it by filaments, which can be traced superficially on the leg.

§ 2. MUSCLES SITUATED ON THE BACK PART OF THE HIP AND THIGH.

These are eleven in number, and may be divided into two classes; 1. Muscles of the hip, and 2. Muscles on the back part of the thigh.

The MUSCLES OF THE HIP are eight in number :

1. The GLUTEUS MAXIMUS—*Arises,* fleshy, from the posterior third of the spine of the os ilium, and from a narrow rough surface of that bone immediately below the spine; from the whole lateral surface of the sacrum, below the ilium ; from the back part of the posterior or great sacro-sciatic ligament *, over which the edge of

---

* The *sacro-sciatic ligaments* are two in number, extended between the ischium and sacrum.

1. The *Posterior, or great sacro-sciatic ligament,* arises from the transverse processes of the sacrum, from the lateral part of that bone, and of the os coccygis ; it is somewhat triangular, descends obliquely, becoming narrower and thicker; and is inserted into the tuberosity of the os ischium. It has here a loose edge, ascending along the ramus of the ischium, which has been called the *falciform ligament.*

2. The *anterior,* or *lesser* ligament is attached, by its base, to the sides of the sacrum, and of the os coccygis ; it passes across on the

this muscle hangs in a folded manner, and from the lateral surface of the os coccygis.

- The fleshy fibres proceed obliquely forwards and downwards, forming a thick broad coarse muscle, and, converging gradually, terminate in a strong flat tendon. This tendon slides over the posterior part of the trochanter major; sends off a great quantity of tendinous fibres, to be inseparably joined to the fascia lata of the thigh; and is

Inserted, into a rough surface at the upper and outer part of the linea aspera, immediately below the trochanter major; also very extensively into the fascia lata, which covers the former insertion.

Situation: It is quite superficial, forming the great mass of the buttock, and covering all the other muscles which are situated on the back part of the hip; covering also by its inferior border the tuber ischii, and the tendons of the muscles which arise from that projection. Its superior border is continuous with the aponeuroses of the sacro-lumbalis and latissimus dorsi. Its insertion lies between the vastus externus and the adductor magnus femoris, and immediately above the origin of the short head of the biceps flexor cruris :—so much adipose membrane is entangled with this muscle, that it is very difficult to dissect it clean, and quite impossible if you do not dissect in the course of the fibres, which are divided into very large and strong fasciculi.

inside of the external, and is inserted, by its apex, into the point of the spinous process of the ischium.

These ligaments convert the large interval, which is found between the lower part of the ilium, spine and tuber of the ischium, into an *upper* and *lower* opening, of which the former gives passage to the pyriformis muscle, the gluteal, ischiatic and pudic vessels, and nerves; the lower opening is filled up by the obturator internus.

Use : To restore the thigh, after it has been bent ;—to rotate it outwards : to extend the pelvis on the thigh, and maintain it in that position in the erect posture of the body.

The muscle is now to be lifted from its origin, and left-hanging by its tendon ; remark the large bursa between the tendon and the trochanter major ; and another between its insertion into the fascia and the tendon of the vastus externus : large branches of the gluteal artery, and also of the ischiatic, and filaments of the gluteal nerves, are seen penetrating the internal surface of the gluteus maximus.

2. The GLUTEUS MEDIUS—*Arises,* fleshy, from all the outer edge of the spine of the os ilium, as far as the posterior tuberosity ; from the dorsum of the bone between the spine, and semicircular ridge, (which passes from the anterior superior spinous process to the ischiatic notch ;) also from the rough surface which extends from the anterior superior to the anterior inferior spinous process, and from the inside of the fascia which covers its anterior part. The fibres converge into a strong and broad tendon, which is

Inserted into the upper and outer part of the great trochanter.

Situation : The posterior part of the belly and tendon are concealed by the gluteus maximus, but the anterior and largest part of this muscle is superficial, being only covered by the fascia, which extends from the front edge of the gluteus maximus to the crista ilii : the anterior border of the gluteus medius lies behind the origin of the tensor vaginæ femoris, and there is, below, an interval filled with cellular tissue between the two muscles :—the posterior border is above the belly of the pyriformis. This muscle covers the gluteus minimus :

there is a bursa between the tendon and the trochanter.

Use : To draw the thigh-bone outwards or away from the opposite limb; by its posterior fibres to rotate the limb outwards, and by its anterior inwards. It acts also in progression, and in maintaining the erect posture.

Having lifted up this muscle from its origin, you will discover one still smaller;

3. The GLUTEUS MINIMUS.—It *arises,* fleshy, from the semicircular ridge of the ilium, and from the dorsum of the bone below the ridge, within half an inch of the acetabulum. Its fibres run in a radiated direction towards a strong tendon, which is

Inserted into the anterior and superior part of the great trochanter.

Situation : It is entirely concealed by the gluteus medius; its internal surface covers the capsule of the hip-joint, and the lower tendon of the rectus femoris: there is a small bursa between the tendon of the gluteus minimus and the trochanter.

Use : The same as that of the preceding.

4. The PYRIFORMIS—*Arises,* within the pelvis, by three tendinous and fleshy origins, from the second, third, and fourth false vertebræ or divisions of the sacrum, between the anterior foramina, which give passage to the nerves. It forms a thick pyramidal belly, which passes out of the pelvis below the niche in the posterior part of the ilium, (from which it receives a few fleshy fibres,) and above the anterior sacro-sciatic ligament.

Inserted, by a roundish tendon, into the uppermost part of the cavity * at the root of the trochanter major.

Situation : In the pelvis it lies behind the rectum, hypogastric vessels, and sciatic plexus : like the other

* *The digital cavity or fossa.*

small muscles of the hip, it is entirely concealed by the gluteus maximus; its belly lies behind and below the gluteus medius, but is not at all covered by it; it crosses from the pelvis to the thigh, along the upper border of the superior gemellus. Its tendon is covered, at the place of insertion, by the posterior fibres of the gluteus medius. Sometimes the pyriformis is divided into two distinct muscles, by a branch of the great sciatic nerve; but generally the nerve comes out from the pelvis below the pyriformis, and above the gemelli: there is a bursa betwixt the tendon of the pyriformis and that of the gluteus medius.

Use: To move the thigh a little upwards, and roll it outwards.

Below the pyriformis, we observe several small muscles lying close together, and passing transversely from the body of the ischium to the root of the great trochanter.

5. The GEMELLI, or GEMINI, consist of two heads, which are distinct muscles:

(1.) The *Superior* arises from the back part of the spinous process of the ischium.

(2.) The *Inferior* from the upper part of the tuberosity of the ischium, and from the anterior surface of the posterior sacro-sciatic ligament. The two muscles pass transversely outwards over the capsule of the hip-joint, and are

Inserted, tendinous and fleshy, into the cavity at the root of the trochanter major, immediately below the insertion of the pyriformis, and above the insertion of the obturator externus.

Situation : Like the other muscles, they are covered by the gluteus maximus; they lie below the inferior border of the pyriformis, and above the quadratus femoris; they are united by a tendinous and fleshy expansion, which forms a purse or sheath for the tendon of the obturator internus: the sciatic nerve descends over the external surface of the gemelli; their internal surface rests on the ischium, and capsule of the hip-joint.

Use: To roll the thigh outwards, and to bind down the tendon of the obturator internus.

Lying between the bellies of the gemelli, you will perceive,

6. The OBTURATOR·INTERNUS.—This muscle *arises* within the cavity of the pelvis; tendinous and fleshy, from the posterior flat surface of the os pubis, within and above the foramen thyroideum; from the internal surface of the obturator ligament; and from the posterior margin of the foramen, and adjacent flattened surface of the ischium, as far back as the sciatic notch. The inner surface of the obturator is covered by a thin fascia, prolonged from the general fascia of the pelvis, and which braces the muscle to the subjacent bone and ligament. The fleshy fibres converge to form a flattened tendon, which passes out of the pelvis, in the lesser notch or sinuosity between the spine and tuberosity of the ischium, then changes its direction, passing outwards over the capsule of the hip-joint, and, becoming rounder, is

Inserted, into the pit at the root of the trochanter major.

Situation: Its origin lies within the pelvis, and cannot be fully exposed until the contents of that cavity are removed; the *sub-pubic,* or *obturator notch* or foramen is then seen, bounded above by the body of the pubis, below by the obturator fascia, muscle and ligament, the fascia forming an inverted arch beneath the vessels and nerve. The tendon, where it passes through the notch in the ischium, is seen projecting between the two origins of the gemelli, and is covered by the gluteus maximus ; but, farther forward, it is enclosed as in a sheath, and concealed by the gemelli, and is inserted between them : it is divided into four or five fasciculi, which impress the pulley-like surface of the ischium, over which they glide, and a loose bursa is interposed. There is also a bursa between the tendon of the pyriformis and the superior gemellus ; and these small muscles are all intimately connected at their insertion.

Use : To roll the os femoris obliquely outwards.

7. The QUADRATUS FEMORIS—*Arises*, tendinous and fleshy, from an oblique ridge, which descends from the inferior edge of the acetabulum along the body of the ischium, between its tuberosity and the foramen thyroideum : its fibres run transversely, to be

Inserted, fleshy, into a rough ridge on the back part of the femur, extending from the root of the greater trochanter to the root of the lesser.

Situation : It is concealed by the gluteus maximus, lies below the inferior head of the gemelli, and above the superior fibres of the adductor magnus : its origin is in contact with the origin of the hamstring muscles : the sciatic nerve also descends over this muscle.

Use : To roll the thigh outwards.

On lifting up the quadratus femoris from its origin, and leaving it suspended by its insertion, you discover, running in the same direction, the strong tendon of,

8. The OBTURATOR EXTERNUS.—This muscle *arises*, fleshy, from almost the whole circumference of the foramen thyroideum, and from the external surface of the obturator ligament ; its fibres pass outwards through the notch between the inferior margin of the acetabulum

and the tuberosity of the ischium, wind round the cervix of the os femoris, adhering to the capsular ligament, and terminate in a strong tendon, which is

Inserted into the lowermost part of the cavity, at the root of the trochanter major, immediately below the insertion of the inferior head of the gemelli.

Situation: This muscle cannot be distinctly seen, until all the muscles which run from the pelvis to the upper part of the thigh are removed, both on the fore and back part; its origin lies on the fore part, filling up the obturator fossa, and concealed by the pectineus and adductor brevis: the obturator vessels come out on its anterior margin. On the back part, the tendon is concealed by the quadratus femoris, and, when that muscle is removed, it is found to run along the lower edge of the inferior head of the gemelli.

Use: To roll the thigh-bone obliquely outwards.

The MUSCLES ON THE BACK PART OF THE THIGH are three in number, springing together from the tuber ischii, forming a large packet on the posterior surface of the adductor magnus, and separating below to pass to opposite sides of the ham. They are bound together by a single cellular sheath, derived from the fascia lata.

9. The BICEPS FLEXOR CRURIS—*Arises*, by two distinct heads; the first, called the *Long Head*, arises in common with the semitendinosus, by a short tendon, from the outer part of the tuberosity of the ischium, and, descending, forms a thick fleshy belly. The second, termed the *Short Head*, arises, tendinous and fleshy, from the linea aspera, immediately below the insertion of the gluteus maximus; and from the oblique ridge running to the outer condyle, where it is connected with the fibres of the vastus externus. The two heads unite

G

at an acute angle a little above the external condyle, and terminate in a strong tendon, which is

Inserted into a rough surface on the outside of the head of the fibula, where it is also connected with the fascia of the leg.

Situation : The long head of this muscle is concealed at its upper part by the inferior fibres of the gluteus maximus : below this, it is situated quite superficial, immediately under the fascia lata, running from the pelvis to the knee between the vastus externus and se-mitendinosus, and forming the outer hamstring.—The short head is partly concealed by the long head, its fibres arise from the linea aspera, between those of the adductor magnus and vastus externus.

Use : To bend the leg, and, by means of its shorter head, to twist the leg outwards in the bent state of the knee.

10. The SEMITENDINOSUS—*Arises,* tendinous, in com-mon with the long head of the biceps, from the tube-rosity of the ischium ; it has also some fleshy fibres arising from that projection more outwardly : as it de-scends, it arises, for two or three inches, fleshy, from the inside of the tendon of the biceps ; it forms a thick belly, and terminates at the distance of three or four inches from the knee in a long round tendon, which de-scends behind the inner condyle, between the inner head of the gastrocnemius and the semimembranosus, and is reflected on the tibia from behind forwards, becoming much broader. It unites with the posterior border of the tendon of the gracilis, and is

Inserted into the inner side of the tibia, some little way below its tubercle.

Situation : This muscle, as well as the biceps, is co-vered above by the gluteus maximus ; its belly lies between the biceps flexor and gracilis, and is situated entirely superficial beneath the fascia lata : its tendon is inserted below that of the gracilis, being closely united with it, and being also connected with the expanded tendon of the sartorious, which passes over them. The belly of this muscle is intersected, about its middle, by a narrow transverse tendinous line.

Use : To bend the leg backwards, and a little inwards.

11. The SEMIMEMBRANOSUS—*Arises,* by a strong round tendon, from the upper and outer part of the tuberosity of the ischium ; the tendon, soon becoming broader, sends off obliquely a fleshy belly ; this muscle is continued fleshy much lower down than that last described.—The fleshy fibres terminate obliquely in another flat tendon, which passes behind the inner condyle, and sends off a thin aponeurotic expansion under the inner head of the gastrocnemius, to cover the posterior part of the capsule of the knee-joint, and to be affixed to the external condyle : the tendon then becoming rounder, is

Inserted into the inner and back part of the head of the tibia, sending off from its posterior edge a strong aponeurosis, which passes over the popliteus muscle.

Situation : This is a semi-penniform muscle ; its origin lies anterior to the tendinous origin of the two last muscles, and at the same time more outwardly, being situated between them and the origin of the quadratus femoris :—its belly, in its descent, is at first concealed by the biceps and semitendinosus ; but, at its lower part, it appears projecting between them. It lies in contact with the posterior surface of the triceps magnus. The tendon, behind the condyle, is anteriorly in close contact with the inner tendon of the gastrocnemius, and there is a synovial capsule interposed.

Use: To bend the leg backwards.

The two last-described muscles properly form the inner hamstring; but some enumerate among the tendons of the inner hamstring, the sartorius and gracilis: there is a bursa between the tendon of the semimembranosus and the head of the tibia; another is observed between the tendons of the gracilis and semitendinosus, and the internal lateral ligament of the knee-joint; and a third is discovered between the tendon of the biceps and the external lateral ligament.

§ 3. OF THE VESSELS AND NERVES ON THE POSTERIOR PART OF THE HIP AND THIGH.

ARTERIES.

1. The GLUTEAL, or POSTERIOR ILIAC ARTERY.— This is the largest branch of the internal iliac artery; it passes out of the pelvis at the upper part of the sciatic notch. On raising the gluteus maximus, and medius, this artery is seen coming over the pyriformis, between the superior edge of that muscle and the inferior edge of the os ilium, (where that bone forms the upper part of the sciatic notch,) and immediately behind the posterior fibres of the gluteus minimus. It immediately divides into two branches;

(*a*,) The *Superficial* branch passes forwards between the gluteus maximus and medius, distributing its ramifications to the substance of both muscles.

(*b*,) The *Deep-seated* branch passes under the gluteus medius, and subdivides; one artery forms an arch along the convex border of the gluteus minimus, while other branches are distributed to that muscle and to the gluteus medius, and to the parts about the hip, and back part of the thigh.

2. The ISCHIATIC ARTERY, is another large branch of the internal iliac, which is seen in the hip on turning back the gluteus maximus : it comes out from under the pyriformis, between the lower edge of that muscle and the superior gemellus. The principal branches of the sciatic artery descend in the fossa between the trochanter major and the tuberosity of the ischium ; it also gives ramifications to the gluteus maximus, and to the parts about the os coccygis and tuber ischii ; and one artery, which seems a continuation of the main trunk, accompanies the sciatic nerve in the back part of the thigh, giving branches to the muscles, and inosculating with the perforating arteries.

Both these arteries inosculate with the other branches of the internal iliac, and with the profunda.

The PUDIC ARTERY is also seen for a short space in the hip, on raising the gluteus maximus,—emerging with the sciatic artery below the pyriformis, then closely applied to the anterior sacro-sciatic ligament, and turning inwards within the greater or posterior sciatic ligament, to reach the perineum.

The VEINS correspond exactly to the arteries. They terminate in the internal iliac vein.

NERVES.

NERVUS ISCHIATICUS, or the GREAT SCIATIC NERVE, is the only important nerve in the back part of the thigh. It is seen coming out of the pelvis along with the sciatic artery, below the pyriformis, and is covered by the lower portion of the gluteus maximus. It descends over the gemelli and quadratus femoris, in the hollow betwixt the great trochanter and the tuberosity of the ischium, and runs down the back part of the thigh, anterior to, that is, nearer the bone than the hamstring muscles; being situated between the anterior surface of the semimembranosus, and the posterior surface of the adductor magnus. It gives branches to these muscles, and also some cutaneous filaments which descend about the ham. Continuing its course along the posterior part of the thigh, the sciatic nerve divides, generally at some short distance above the ham, into two branches; the *Peroneal Nerve*, and the *Popliteal* or *Posterior Tibial Nerve*, which last appears the continued trunk. The sciatic sometimes perforates the belly of the pyriformis by distinct branches, which afterwards unite.

Besides this great nerve, when the gluteus maximus is raised, the smaller posterior branches of the Crural Plexus are exposed. The *Gluteal nerve* is seen coming out above the pyriformis, to ramify in the gluteal muscle, extending as far as the tensor muscle of the fascia. The *lesser sciatic,* or *inferior gluteal* nerve, comes out below the pyriformis, and is seen in the hip posterior to the great nerve; it gives *(a,) Gluteal branches* to the muscles of that name; *(b,) A sciatic branch,* which descends below the tuber ischii to the internal part of the thigh and perineum, *(c,)* the long *posterior cutaneous branch,* before noticed. The *Pudic* nerve corresponds with its artery.

SECTION III.

DISSECTION OF THE HAM.

On removing the integuments from the back part of the knee-joint, we observe a FASCIA, thin but firm, which covers the great vessels and the muscles. It is evidently continued from the great fascia of the thigh, is connected by adhesions to the condyles of the femur, and to the head of the fibula, and is prolonged upon the muscles on the back of the leg.

The Ham is the triangular, or oval hollow, formed at this part; the fascia stretches over it; it is bounded, laterally, by the tendons of the semitendinosus and semimembranosus on the inner side, and by the biceps on the outer side, and, lower down, by the two heads of the gastrocnemius.

Upon dissecting back the fascia covering the ham, a considerable quantity of soft fat is observed, some of which being removed, the GREAT SCIATIC NERVE first appears, lying between the outer and inner hamstring muscles. It is seen dividing, at some distance above the condyles of the femur, into its two branches.

1. The greater nerve continues its course behind the articulation of the knee, and betwixt the heads of the gastrocnemius muscle. In the ham it is named the POPLITEAL NERVE, and where it descends in the leg, POSTERIOR TIBIAL: about an inch above the condyle, the popliteal nerve gives off a *Cutaneous branch*, which descends beneath the fascia, in the interval between the two portions of the gastrocnemius, with the *external* or *minor saphena* vein; this superficial branch gives filaments to the integuments, and unites with a branch of the Peroneal nerve to form a trunk, which, ramifying on the outer side of the leg and foot, has been called the *Nervus Saphenus Externus*.

2. The lesser Nerve, which is the external branch, is named the PERONEAL or FIBULAR NERVE; it passes outwards and obliquely downwards, runs between the external head of the gastrocnemius, and the tendon of

the biceps flexor cruris, and sinks among the muscles which surround the head of the fibula, where it divides into two branches, the *Musculo-cutaneous*, and the *Anterior Tibial.*

But the Peroneal nerve also gives off above the condyles of the femur a *Cutaneous branch*, which descends over the gastrocnemius along the back of the leg, under the fascia, and is distributed to the integuments and to the outer side of the foot : it sends a filament to form the Nervus Saphenus externus.

The terminating branches of the Peroneal nerve bend forwards to the fore part of the leg, viz. :

(1.) The *Musculo-cutaneous Nerve* descends between the peronei and extensor longus digitorum, and afterwards becomes superficial; it is seen in the dissection of the fore part of the leg.

(2.) The *Anterior Tibial Nerve* passes under the upper fleshy extremities of the peroneus longus and extensor digitorum longus, supplying them, and comes in contact with the anterior tibial artery, which it accompanies down the leg.

Beneath the great sciatic nerve, there is much cellular membrane and fat, and three or four lymphatic glands, which being removed, the GREAT POPLITEAL VEIN is exposed. It adheres to the POPLITEAL ARTERY, which lies under it close upon the bone.

The POPLITEAL ARTERY is the trunk of the FEMORAL, which assumes that name after it has perforated the tendon of the triceps. It lies deep-seated between the condyles of the femur, close upon the flat posterior surface of the bone, and ligaments of the knee-joint, and then descends between the heads of the gastrocnemius. It runs over the popliteus muscle, and under the gastrocnemius, that is, in the erect position it is anterior to the gastrocnemius, and posterior to the popliteus: at the lower edge of the popliteus, the popliteal artery divides into the ANTERIOR and POSTERIOR TIBIAL ARTERIES, of which the former passes immediately, through the inter-osseous ligament, to the fore part of the leg, while the latter is continued along the back of the limb.

The branches of the Popliteal artery are,

1. Small branches to the Sciatic nerve and parts about the ham. . .

2. The *Articular Arteries,* usually five in number; ramifying over the knee-joint; and inosculating with one another, with the descending branches of the femoral, and with the recurrent branches below the knee. They are named according to their situation, *superior external, superior internal, inferior external, inferior internal,* and *A. articularis azygos,* or *media.*

3. The *Sural* Arteries, two arteries sent to the heads of the gastrocnemius and plantaris muscles.

The POPLITEAL VEIN receives branches corresponding with those of the artery; it lies behind the artery in the erect posture, and somewhat external to it, and divides into the *Anterior* and *Posterior Tibial* veins: in the hollow of the ham, just above the condyles, it also receives a large subcutaneous branch, the *Vena Saphena Minor* or *Externa,* which ascends from the back part of the leg and outer side of the foot: the trunk of this vein lies under the fascia.

CHAPTER V.

DISSECTION OF THE LEG AND FOOT.

SECTION I.

OF THE FORE PART OF THE LEG AND FOOT.

§ 1. OF THE FASCIA OF THE LEG.

ON dissecting off the integuments from the fore part of the leg, we find a strong fascia continued from the thigh; it is of less thickness, but is evidently prolonged from the fascia lata; it also receives fibres from the tendons of the sartorius, gracilis, and semimembranosus, and from the tendinous expansions of the rectus and vasti ;— it adheres firmly to the projecting points of the bones, to the heads of the tibia and fibula, and, as it descends, to the anterior and inner edges of the tibia, blending inseparably with the periosteum on the internal subcutaneous surface of the latter bone. It invests the whole leg, binding down the muscles, but its greatest thickness is on the fore part of the limb; at the back part of the leg and below, it is much thinner ;—where it passes over the ankle, it again becomes very strong by its adhesions about the outer and inner malleolus; and it here forms the *Anterior Annular Ligament*, which is evidently but a thicker and stronger part of the general fascia of the leg. Below, and on the fore part, the fascia of the leg terminates by a thin tendinous expansion, which covers the dorsum of the foot: on the back part of the leg, it is lost insensibly on the heel; on the outer side it is connected with the sheath of the peronei muscles, and

7

on the inside it is fixed into the *Internal Annular Liga-ment.*

The ANTERIOR ANNULAR LIGAMENT seems to consist of two distinct bands, which, going from the point of the outer ankle and neighbouring part of the os calcis, are fixed, in a wide expanded manner, to the malleolus in-ternus, and to the inside of the os naviculare, where the fibres are connected with the inner edge of the Plantar fascia. This ligament binds down the extensor tendons, forming distinct rings, or sheaths, through which these tendons pass, and the anterior tibial vessels and nerve also pass beneath it.

The INTERNAL ANNULAR LIGAMENT is a broad tendi-nous band, also connected with the fascia of the leg, which passes from the fore part of the inner ankle to the posterior and inner part of the os calcis, stretching across the sheaths of the flexor muscles and over the posterior tibial vessels and nerve.

More superficial than the fascia of the leg, between it and the common integuments, it will be proper to re-mark,

1. The *Vena Saphena Major,* commencing from veins on the inner side and fore part of the foot; it crosses over the inner ankle, and then running upwards upon the inside of the tibia, ascends behind the inner condyle, to join the femoral vein in the groin.

2. The Saphena vein, in its course on the inner side of the knee and down the leg, is accompanied by a cu-taneous nerve, the *Nervus Saphenus Internus :* this is a branch of the anterior crural nerve, and has been already noticed in the dissection of the thigh : it distributes branches superficially, and descends as far as the great toe.

3. Several other *Cutaneous Nerves* are also observed,

branching on the outer and fore part of the leg and foot : they are mostly derived from the branches of the *Peroneal Nerve.* The *Nervus Saphenus Externus*, which is chiefly the continued trunk of the *Cutaneous branch* of the *Popliteal nerve*, is also seen coming from the back part of the leg, where it descends along the external side of the tendo Achillis : it passes behind the external malleolus, and, appearing on the dorsum of the foot, reaches the posterior extremity of the fifth metatarsal bone ; it here divides into *(a,)* an *external* branch to the outer side of the little toe, and *(b,)* an *internal* branch, which supplies the opposed sides of the two last toes.

The fascia should now be dissected off; and, in doing this, remark, that it is firmly attached to the bones, and also to the bellies of the muscles at the upper part of the leg, so that their surfaces appear ragged, where the fibres are separated which arose from the inside of the fascia. Remark also that it sends down processes between the muscles ; these are named intermuscular ligaments, or tendons ; they give origin to the fibres of the muscles betwixt which they pass, connecting them together inseparably, so that the dissection is difficult, and has a rough appearance.

§ 2. MUSCLES SITUATED ON THE FORE PART AND OUTSIDE OF THE LEG.

These are six in number.

I. The TIBIALIS ANTICUS—*Arises*, principally fleshy, from the exterior flattened surface of the tibia, beginning immediately below the head or tuberosity ; *from* its anterior angle or spine, and from nearly half of the interosseous ligament ; from these surfaces it continues to arise down two-thirds of the length of the bone ; also from the inner surface of the fascia of the leg, and from the intermuscular ligament, which connects it with the next muscle. The fleshy fibres descend obliquely, and terminate in a strong tendon, which crosses from the outside to the fore part of the tibia, passes through a distinct ring of the annular ligament near the inner ankle, runs over the astragalus and os naviculare to the inner edge of the foot, and, becoming broader, is

Inserted, by a bifurcated extremity, into the upper and inner part of the os cuneiforme internum, and into the base of the metatarsal bone supporting the great toe.

Situation: The belly of this muscle is placed on the outer or fibular side of the tibia, and is quite superficial, lying under the fascia of the leg : on its outer border are the two next muscles, and the anterior tibial vessels and nerve : the insertion of the tendon is concealed in part by the adductor and flexor brevis of the great toe. Between the tendon of this muscle and the os cuneiforme, we find a small bursa mucosa ; and, in passing through the ring of the annular ligament, the tendon has a synovial capsule.

Use: To draw the foot upwards and inwards; or, in other words, to bend the ankle joint. Also to keep the leg perpendicular on the foot, when the fixed point is below.

2. Extensor Longus Digitorum Pedis—*Arises,* tendinous and fleshy, from the outer part of the head of the tibia ; from the head of the fibula ; from the anterior angle of the fibula almost its whole length ; and from part of the smooth surface between the anterior and internal angles ; also from a small part of the interosseous ligament ; and from the fascia and intermuscular ligaments. The fleshy fibres end, below the middle of the leg, in four round tendons, which pass under the annular ligament, become flattened, and are

Inserted into the root of the first phalanx of each of the four small toes, being connected with the inner borders of the tendons of the extensor brevis; the united tendons are expanded over the upper surface of the toes as far as the root of the last phalanx.

Situation : This muscle also runs entirely superficial; it lies between the tibialis anticus and peroneus longus, arising between them, and being connected to them by intermuscular ligaments; but, at the middle part of the leg, it is separated from the tibialis anticus by the extensor pollicis longus, and from the peroneus longus by the peroneus brevis : nearer the ankle, its outer border is connected with the next muscle : where the tendons pass beneath the annular ligament, they have a synovial sheath.

Use : To extend all the joints of the four small toes : —to bend the ankle joint.

3. PERONEUS TERTIUS—*Arises*, fleshy, from the anterior angle of the fibula, and from part of the smooth surface between the anterior and internal angles, extending from below the middle of the bone downwards to near its inferior extremity; it sends its fleshy fibres forwards to a tendon, which passes under the annular ligament, in the same sheath with the tendons of the extensor digitorum longus, and, becoming broad, is

Inserted into the outer side of the base of the metatarsal bone that supports the little toe.

Situation : The belly is inseparably connected with the extensor longus digitorum, and is properly the outer part of it; it lies between that muscle and the peroneus brevis. The tendon runs down on the outside of that tendon of the extensor longus digitorum, which goes to the little toe. The whole of the muscle is superficial : it is sometimes wanting, and also varies in size.

Use : To assist in bending the foot, and to incline it outwards.

4. EXTENSOR PROPRIUS POLLICIS PEDIS, or EXTENSOR POLLICIS LONGUS—*Arises*, tendinous and fleshy, from part of the smooth surface between the anterior and internal angles of the fibula, and from the neighbouring part of the interosseous ligament, extending from some distance below the head of the bone to near its inferior extremity ; a few fibres also arise from the lower part of the tibia ;—the fibres pass obliquely downwards and forwards into a tendon, which, inclining inwards, passes under the annular ligament in a separate ring or passage, and runs on the inner side of the dorsum of the foot, to be

Inserted into the base of the first and of the second phalanges of the great toe *.

Situation : The belly is concealed between the tibialis anticus and extensor digitorum longus, and cannot be seen till those muscles are separated from one another ; —the tendon is superficial, running between the tendons of those two muscles, and in its passage under the annular ligament has a synovial capsule.

Use : To extend the great toe ; and to bend the ankle.

5. The PERONEUS LONGUS—*Arises*, tendinous and fleshy, from the fore part and outside of the head of the fibula, and from the adjacent part of the tibia : from the external angle of the fibula, and from the smooth surface between the anterior and external angles, as far down as one third of the length of the bone from its lower extremity ; also from the fascia of the leg, and intermuscular ligaments, which connect this muscle with the soleus and flexor longus pollicis on one side, and

* N. B. It is to be understood, that the great toe has only two phalanges.

with the extensor digitorum longus on the other. The fibres run obliquely outwards into a tendon, which passes behind the outer ankle, through a groove in the lower extremity of the fibula, which is common to it and to the tendon of the next muscle; it is here tied down by a ligament, and is invested with a synovial capsule: it is then reflected forwards through a groove in the outer side of the os calcis, where it is also bound down and is invested with synovial membrane; it then turns over the side of the os cubojdes, enters the deep groove in the under part of that bone, where it has its ligamentous and synovial sheath; and runs obliquely inwards across the muscles in the sole of the foot, to be

Inserted into the outside of the base of the metatarsal bone that sustains the great toe, and into the os cuneiforme internum.

Situation: The belly is quite superficial under the fascia; it lies between the outer edge of the extensor longus digitorum, and the anterior edge of the soleus, being connected to both. The tendon is superficial where it crosses the outside of the os calcis, but, in the sole of the foot, is concealed by the muscles situated there, and will be seen in the dissection of that part: there is generally a sesamoid bone in the tendon, where it passes over the side of the os cuboides.

Use: To extend the ankle joint, turning the sole of the foot outwards.

6. The PERONEUS BREVIS—*Arises*, fleshy, from the outer edge of the anterior angle of the fibula, and from part of the smooth surface behind that angle, beginning about one-third down the bone, and continuing its adhesion to near the ankle; from the fascia of the leg, and from the intermuscular ligaments, which separate it, anteriorly, from the peroneus tertius and extensor longus

digitorum, and, posteriorly, from the flexor longus pollicis.—The fibres run obliquely towards a tendon, which passes through the groove of the fibula behind the outer ankle, being there enclosed in the same ligament with the tendon of the peroneus longus, then through a separate groove on the outside of the os calcis, and is

Inserted into the external part of the base of the metatarsal bone that sustains the little toe.

Situation: This muscle arises between the extensor longus digitorum and peroneus longus; its belly is overlapped, and concealed by the belly of the peroneus longus; but as it continues fleshy lower down, it is seen, above the ankle, projecting on each side of the tendon of that muscle. Below, it is separated from the peroneus tertius by that projection of the fibula which forms the outer ankle, and which is only covered by the common integuments, and its posterior edge here ranges with the flexor longus pollicis. The tendon, where it passes through the groove of the fibula, lies under that of the peroneus longus, *i. e.* nearer the bone, but it is soon seen before it, and, on the side of the os calcis, runs above it.

Use: The same as that of the peroneus longus.

§ 3. MUSCLES ON THE UPPER PART OF THE FOOT.

Only one muscle is found in this situation; it is covered by the thin aponeurosis, which passes from the fascia of the leg over the dorsum of the foot.

EXTENSOR BREVIS DIGITORUM PEDIS—*Arises*, fleshy and tendinous, from the anterior and upper part of the os calcis, from the os cuboides, and from the astragalus; it forms a fleshy belly, divisible into four portions: these send off four slender tendons, which proceed forwards over the metatarsal bones, crossing the tendons of the extensor longus.

Inserted, by its first tendon, into the base of the first phalanx of the great toe, lying under the tendon of the extensor longus, and, by the three remaining tendons, into the next three toes, here lying on the outer side, and uniting with the tendons of the long extensor.

Situation : The belly of this muscle lies under the tendons of the extensor digitorum longus and peroneus brevis : it is not, however, concealed, but is seen projecting behind and betwixt these tendons ; it assists in forming the tendinous membrane, which invests the upper surface of all the phalanges of the toes.

Use : To extend the toes.

It is sometimes described as two muscles, the extensor brevis pollicis pedis, and extensor brevis digitorum pedis.

§ 4. OF THE VESSELS AND NERVES IN THE FORE PART OF THE LEG AND FOOT.

I. ARTERIES.

The ANTERIOR TIBIAL ARTERY passes from the ham betwixt the inferior edge of the popliteus, and the superior fibres of the soleus, and then through a large perforation in the interosseous ligament, to reach the fore part of the leg ; this perforation is much larger than the artery itself, and is filled up by the fibres of the tibialis posticus muscle, which may thus be said to arise from the fore part of the tibia, and, having its origin from both bones, appears to bifurcate to give passage to the vessels. The artery then runs down close upon the middle of the interosseous ligament, first, between the tibialis anticus and extensor longus digitorum, and, lower down, between the tibialis anticus and extensor proprius pollicis : below the middle of the leg, it leaves the interosseous ligament, and advances gradually more forwards. It now takes its course upon the fore part of the tibia, becoming more. superficial, and passes beneath the annular ligament, and it is here crossed by the tendon of the extensor proprius pollicis. Having passed the ankle-joint, it is continued directly forwards upon the tarsus, being here quite superficial and somewhat tortuous, and lying between the tendon of the extensor pollicis and the first tendon of the extensor longus digitorum ; it runs over the astragalus, navicular, and internal cuneiform bones, crossing under that tendon of the extensor brevis digitorum which goes to the great toe ; then, arriving at the space between the bases of the two first metatarsal bones, it plunges into the sole of the foot, and immediately joins the plantar arch.—In its course down the leg, the anterior tibial artery is accompanied by two veins, and by the anterior tibial nerve which lies on its fore part.

BRANCHES OF THE ANTERIOR TIBIAL ARTERY.

1. *A. Recurrens* comes off from the anterior tibial after it has passed through the interosseous ligament ; it ascends in the substance of the upper part of the tibialis anticus, to which it gives many branches ; then perforating the fascia of the leg, it ramifies around the knee, inosculating with the lower articular arteries.

2. Numerous *Muscular* twigs to the tibialis anticus, extensor pollicis, and other muscles on the fore part of the leg.

3. *A. Malleolaris Interna*, passing behind the tendon of the tibialis anticus, ramifies over the inner ankle, and inosculates with the peroneal and posterior tibial arteries.

4. The *External Malleolar* runs behind the tendons of the extensor digitorum longus and peroneus tertius, and ramifies over the outer ankle.

5. and 6. The *Tarsal and Metatarsal Arteries* are two small branches which cross the tarsal and metatarsal bones, and pass obliquely to the outer edge of the foot.

From the tarsal or metatarsal artery come off the *Interosseous Arteries*, which supply the interosseous spaces, and the back part of the toes.

7. A large branch comes off from the anterior tibial, where it is about to plunge into the sole of the foot; it runs along the space betwixt the two first metatarsal bones, and, at the anterior extremity of those bones, bifurcates into

(1.) *A. Dorsalis Hallucis* a considerable branch, which runs on the back part of the great toe.

(2.) A *branch* which runs on the inner edge of the toe next to the great one.

8. *Ramus Anastomoticus* is the short terminating branch, which plunges into the sole of the foot.

II. VEINS.

The ANTERIOR TIBIAL VEINS, two in number, are placed on each side of the artery; at the upper part of the leg, they unite to form a single trunk, which passes through the interosseous ligament to join the Popliteal Vein.

III. NERVES.

1. The ANTERIOR TIBIAL NERVE is a branch of the peroneal nerve, coming off from that nerve while it is crossing round the fibula under the head of the peroneus longus. The anterior tibial nerve continues round

to the fore part of the leg, sending filaments to the adjacent muscles, and emerges from the extensor longus digitorum, to which muscle also it gives a considerable branch : it soon comes in contact with the anterior tibial artery, and accompanies it down the leg, lying in front of the artery : it passes under the annular ligament with the tendon of the extensor pollicis proprius, and divides on the dorsum of the foot into two branches : (1.) The *internal* passes forwards, under that portion of the extensor digitorum brevis which goes to the great toe, and runs between the two first metatarsal bones, supplying the muscles and integuments, and the inner and outer side of the two first toes. (2.) The *external* branch sinks beneath the extensor digitorum brevis, and subdivides into numerous filaments to that muscle and the interossei.

2. The MUSCULO-CUTANEOUS NERVE, the second terminating branch of the peroneal, is also seen in this dissection ; it is found deep between the peronei and extensor longus digitorum, and here gives off muscular branches. About the middle of the leg, it is less deeply seated, descending for some distance beneath the fascia, which it then perforates, and, becoming sub-cutaneous, continues its course downwards : having given some filaments to the integuments of the outer ankle, it divides into two branches, of which (*a,*) The *internal* ramifies on the inner border of the foot and upper surface of the first and second toes. (*b,*) The *external* passes on the dorsum of the foot, between the extensor tendons and the integuments, and divides into three branches, which, passing to the intervals between the four outer toes, subdivide, and supply their opposite surfaces, constantly anastomosing with the nervus saphenus externus.

SECTION II.

DISSECTION OF THE POSTERIOR PART OF THE LEG.

§ 1. OF THE FASCIA.

THE fascia which invests the posterior part of the leg, is much thinner and less strong than on the fore part : it is a continuation of that portion of the general fascia which covers the ham, and is lost below in the fibro-cellular tissue about the heel : it must be removed, to expose the parts now to be described.

The *Vena Saphéna Minor,* or *Externa;* begins by cutaneous branches on the outer ankle and side of the foot : it is seen ascending from the outer ankle, along the middle of the gastrocnemius muscle, to join the popliteal vein ; at first it is sub-cutaneous, but in the upper part of the leg it gradually sinks between the laminæ of the fascia, to enter the hollow of the ham. The *cutaneous branches* of the popliteal and peroneal nerves, described in the dissection of the ham, are also seen : below the middle of the leg, the *external saphenous nerve* is closely connected with the lesser saphena vein, descending with it behind the fibula, and then turning behind the external malleolus to the dorsum of the foot.

§ 2. MUSCLES ON THE POSTERIOR PART OF THE LEG.

These are seven in number, and may be divided into two classes.

MUSCLES situated more superficially,—four in number.

1. The GASTROCNEMIUS EXTERNUS, or GEMELLUS— *Arises,* by two distinct heads ;—The *first,* or *Internal Head,* arises, tendinous, from the upper and back part of the internal condyle of the os femoris, and, fleshy,

from the oblique ridge above that condyle. The *second,* or *External Head,* arises in the same manner, from the external condyle. Each of the heads forms a fleshy belly, the fibres of which are oblique, passing from a tendinous expansion which covers the posterior surface of the muscle, to another tendinous expansion which covers the anterior surface, or that surface which lies nearest the bones. The two bellies, of which the internal is by much the largest, are separated by a considerable triangular interval, in which the popliteal blood-vessels and nerves pass to the leg; but descending, they unite a little below the knee-joint in a middle tendinous line; and, below the middle of the tibia, send off a broad flat tendon, which unites a little above the ankle with the tendon of the soleus.

Reflect the two heads of the gastrocnemius from the femoral condyles, and you will then expose

2. The SOLEUS or GASTROCNEMIUS INTERNUS—which also *arises,* by two origins or heads. The *first,* or *External* origin, which is by much the largest, arises, principally fleshy, from the posterior surface of the head of the fibula, and from the external angle or margin of that bone, for two-thirds of its length, immediately behind the peroneus longus. The *second,* or *Internal* head, *arises,* fleshy, from an oblique ridge on the posterior surface of the tibia, just below the popliteus, and from the inner angle of that bone, during the middle third of its length. The two heads are at first separated by the popliteal vessels, but they are soon united by their inner borders, forming a *fibrous arch,* with its concavity upwards, over the vessels. The fleshy fibres descend obliquely, and terminate in a thin aponeurosis, which covers the posterior surface of the muscle. Below, this aponeurosis unites with the tendon of the gastrocnemius,

to form a strong round tendon, named the *Tendo Achillis*, which slides over a smooth cartilaginous surface on the upper and posterior part of the os calcis, to be

Inserted into a rough surface on the lower and back part of that bone.

Situation : These two powerful muscles, at first separated by a cellular interval, finish by blending inseparably their tendons, and are described by some as a *musculus triceps* of the calf of the leg. The gastrocnemius arises between the hamstring tendons : its belly is superficial, and forms the upper or greater calf : on raising it, the tendon is seen continued some way on its inner surface.

The soleus has its larger part concealed by the gastrocnemius, but part of it appears on each side of the belly of that muscle ; and, at the lower part of the leg, the belly is seen projecting through the tendon of the gastrocnemius, and forming the lower calf. The tendo Achillis is immediately beneath the skin ; it is separated from the deep flexor muscles by much adipose tissue, and it receives fleshy fibres from the soleus to near the heel : there is a bursa between the tendon and the upper part of the os calcis.

Use : To elevate the os calcis, and thereby, to lift up the whole body, as a preparatory measure to its being carried forward in progression ;—to carry the leg backwards on the foot when that is fixed ;—the gastrocnemius, from its origin in the thigh, also bends the leg on the thigh.

The external head of the gastrocnemius being reflected from the condyle, we expose

3. The PLANTARIS.—This muscle *arises*, fleshy, from the upper part of the external condyle, and from the oblique ridge above that condyle ; it forms a pyramidal

belly about three inches in length, which adheres to the capsule of the knee-joint, runs over the popliteus, and terminates in a long, slender, thin tendon. This tendon passes obliquely inwards over the inner head of the soleus, and under the gastrocnemius; emerges from between the two muscles, where their tendons unite, and then runs down close to the inner side of the tendo Achillis, to be

Inserted into the posterior part of the os calcis, on the inside of the insertion of the tendo Achillis, and somewhat before it.

Situation : The origin and belly of this muscle are concealed by the external head of the gastrocnemius : the lower part of the tendon is the only part that is superficial : at its insertion, it is closely united to the tendo Achillis. This muscle is sometimes wanting.

Use : To extend the foot, and roll it inwards, and to assist in bending the leg.

Immediately behind the knee, and above the soleus, we observe a small triangular muscle, covered by a fascia;

4. The POPLITEUS—*Arises,* within the ligaments of the articulation of the knee, by a strong flattened tendon, from a deep pit or hollow on the outer side of the external condyle. This tendon adheres to the posterior and outer surface of the external semilunar cartilage, and passes over the side of the condyle to its back part; it is placed exterior to the synovial cavity of the articulation, but is invested by a process or reflexion of the synovial membrane; it glides above the upper peroneo-tibial articulation, and forms a fleshy belly, which descends obliquely inwards, covered by a thin tendinous fascia, to be

Inserted, broad, thin, and fleshy, into an oblique

H

ridge on the posterior surface of the tibia, a little below its head, and into the triangular space above that ridge.

Situation: Its origin is concealed by the external lateral ligament, and tendon of the biceps flexor: the muscle itself crosses immediately behind the articulation of the knee, and is concealed entirely by the gastrocnemius, lying above the the inner head of the soleus; it is more deeply seated than the plantaris, which crosses over it; its fascia seems derived from the tendon of the semimembranosus.

Use: To bend the leg, and, when bent, to roll it, so as to turn the toes inwards.

The belly of the soleus should now be lifted, in order to expose the deeply-seated muscles; connecting and investing these muscles, a strong *Fascia* is seen, extending across from the inner and posterior edge of the tibia to the outer edge of the fibula, and continued, beneath the tendo Achillis, from the general fascia of the leg: this deep-seated fascia is of greatest thickness below; as it ascends over the flexor muscles, beneath the soleus, it becomes thinner and less marked. It also covers the posterior tibial vessels and nerve, but the description of the course of these vessels, though seen in this stage of the dissection, must be deferred.

The deep-seated muscles are,

The Flexor Longus Digitorum Pedis, situated behind the tibia.

The Flexor Longus Pollicis Pedis, situated behind the fibula.

The Tibialis Posticus, which is almost concealed by the two other muscles, and by the fascia, which connects them, and binds them down; and which must now be removed.

5. The FLEXOR LONGUS DIGITORUM PEDIS PERFO-

RANS—*Arises*, fleshy, from the posterior flattened surface of the tibia, below the attachment of the soleus, and from the internal and external angles or edges of that bone, by two distinct orders of fibres, between which the tibialis posticus is seen enclosed: it contiues to arise from the bone to within two or three inches of the ankle; the fibres run obliquely into a tendon, which is situated on the posterior edge of the muscle. This tendon runs behind the inner ankle in a groove of the tibia, which is common to it with the tendon of the tibialis posticus, and passes deep, beneath the astragalus, into the sole of the foot: it then continues forwards, and, having received a strong tendinous slip from the flexor pollicis longus, divides about the middle of the sole of the foot into four tendons, which pass through the slits in the tendon of the flexor digitorum brevis, and are

Inserted into the last phalanx of each of the four lesser toes.

Situation: The belly of this muscle is concealed by the soleus, and lies on the inside of the flexor longus pollicis, to which and to the tibialis posticus it is united by the common fascia. The situation of the tendon will be described with the muscles situated in the sole of the foot.

Use: To bend the last joint of the toes, and to assist in extending the foot.

6. FLEXOR LONGUS POLLICIS PEDIS—*Arises*, fleshy, from the posterior flat surface of the fibula, continuing its origin from some distance below the head of the bone, to within an inch of the ankle: also from the interosseous ligament, and from the intermuscular ligaments, which separate it from the two peronei on one side, and, on the other, from the flexor digitorum and tibialis posticus. The fleshy fibres terminate in a tendon, which

passes behind the inner ankle, through a groove in the
tibia, and then turns forwards, through a groove in the
astragalus, into the sole of the foot: it is next continued
through a groove in the os calcis, crosses the tendon of
the flexor longus digitorum, to which it gives a slip of
tendon, and directs its course on the inner side of the
foot, between the two portions of the flexor brevis pollicis: it passes between the two sesamoid bones, and is

Inserted into the last phalanx of the great toe.

Situation: The belly of this muscle is covered by the
soleus: it lies on the outer side of the flexor longus digitorum, between that muscle and the peronei muscles;
the tendon, where it passes behind the inner ankle, is
situated more backward than the tendon of the flexor
digitorum longus, that is, nearer the os calcis: its further course will be seen in the dissection of the sole of
the foot.

Use : To bend the last joint of the great toe, and,
being connected by the cross slip to the flexor digitorum
communis, to assist in bending the other toes.

7. The TIBIALIS POSTICUS—*Arises*, fleshy, from the
posterior surface of both the tibia and fibula, immediately below the upper articulation of these bones with
each other: from the greater part of the interosseous
ligament; from the angles of the bones to which that
ligament is attached; and from the flat surface of the
fibula behind its internal angle for more than two-thirds
of its length. The fibres run obliquely towards a middle
tendon, which, becoming round, passes behind the inner
ankle through the groove in the tibia, with the tendon
of the flexor digitorum, and is then continued forwards
into the sole of the foot.

Inserted into the upper and inner part of the os naviculare, being further continued to the internal and ex-

ternal cuneiform bones ; it also sends some tendinous filaments to the os calcis, the os cuboides, and the bases of the metatarsal bones supporting the second and middle toe..

Situation : This muscle may be said to arise from the tibia and fibula before the interosseous ligament, as its fibres fill up a perforation in the upper extremity of that ligament : this origin from the two bones has a bifurcated appearance, giving passage to the anterior tibial vessels. The belly is concealed at its lower part by the flexor longus digitorum and flexor pollicis, but a part of it is discovered above the upper extremity of these muscles, and immediately below the fibres of the popliteus ; and this part lies under the anterior surface of the soleus. The tendon crosses under that of the flexor longus digitorum above the ankle, and, where it passes through the groove in the tibia, is situated more forward than the tendon of that muscle, lying immediately behind the malleolus. It generally has a sesamoid bone, where it passes beneath the astragalus : and its insertion lies close in contact with the bones, and is concealed by the muscles in the sole of the foot.

Use : To extend the foot and turn it inwards : it also supports the trochlea, or strong fibro-cartilage, on which the head of the astragalus rests.

The internal Annular Ligament has been formerly described, as stretching from the inner ankle to the os calcis over the flexor tendons ; it forms, with the sinuosity of the os calcis, a kind of channel, through which they pass : the tendons are tied down by ligamentous sheaths to the grooves through which they glide, and they are invested by sheaths or capsules of synovial membrane : there is also a considerable quantity of fatty substance in the hollow between the inner ankle and heel.

§ 3. OF THE VESSELS AND NERVES OF THE POSTERIOR PART OF THE LEG..

I. ARTERIES.

The POSTERIOR TIBIAL ARTERY, which is the conti-nued trunk of the popliteal, sinks under the tendinous arch connecting the two origins of the soleus, and runs down the leg, between that muscle and the more deeply-seated flexor muscles : it lies close upon the deep-seated muscles, immediately under the fascia which binds them down, together with its veins and accompanying nerve. Its direction is in a line from the middle of the ham to the hollow behind the inner ankle; above, it is deep-seated, taking its course over the tibialis posticus and flexor longus digitorum, and being covered by the soleus and gastrocnemius : as it descends, it gradually advances more forwards, emerges from beneath the soleus, and is placed on the inner edge of the tendo Achillis, taking its course downwards over the flexor digitorum longus, and at the distance of less than an inch from the inner margin of the tibia; it is here more superficial, being only covered by the fascia and integuments. It follows the direction of the flexor tendons, passes behind the inner ankle, in the sinuosity of the os calcis, lying pos-terior to the tendons of the tibialis posticus and flexor longus digitorum, and anterior to that of the flexor lon-gus pollicis. Here it is close upon the bone, and its pulsation may be felt about an inch from the malleolus internus. It sinks under * the abductor pollicis arising from the os calcis, and immediately divides into the *In-ternal* and *External Plantar* arteries.

The branches of the posterior tibial artery are,

1. The *Peroneal Artery*, which comes off from the

---

* The sole of the foot is supposed to be turned uppermost.

tibial an inch or two below the origin of the anterior
tibial. This artery is generally of a considerable size,
sometimes nearly as large as the tibial itself; it is situated deeply at the back part of the leg, descending for
a short distance nearly parallel to the posterior tibial,
then taking its course along the inner and back part of
the fibula to near the outer ankle ;—it is at first situated
upon the tibialis posticus, and then descends between
that muscle and the flexor longus pollicis; it sinks under
the fibres of the flexor longus pollicis, and, having given
off branches to it and to the adjacent muscles, terminates below the middle of the leg, by dividing into an
anterior and posterior branch :

(*a,*) The *Anterior Peroneal Artery* passes through the
interosseous ligament to the fore part of the leg, runs
under the peroneus tertius, and over the articulation
of the tibia with the fibula, to the back part of the foot.

(*b,*) The *Posterior Peroneal Artery* is properly the termination of the artery, following its original direction,
and descending behind the external malleolus to the
outer side of the os calcis; it gives branches to the
neighbouring parts, and terminates on the outer and
back part of the foot and muscles of the little toe.

2. *Muscular branches* are given off by the tibial artery in its descent down the leg, and an artery, the *Nutritia tibiæ*, to the tibia itself; and branches pass off
from the tibial, where it passes in the hollow of the os
calcis, to the muscles of the foot and integuments.

The terminating branches of the posterior tibial are
the two *Plantar* arteries, which will be seen in the dissection of the foot, but may be here described.

(1.) The *Internal Plantar Artery* is the smallest, and
ramifies among the mass of muscles situated on the inner
edge of the sole of the foot.

(2.) The *External Plantar Artery* is larger, and appears the continued trunk of the posterior tibial.. It directs its course obliquely outwards, passing between the flexor brevis digitorum and the flexor accessorius, giving numerous muscular branches; having reached the base of the metatarsal bone of the little toe, it bends inwards across the foot, and forms the *Plantar Arch:* This arch crosses the three middle metatarsal bones obliquely, about their middle, lying deep under the flexor tendons, and terminates at the space betwixt the two first metatarsal bones, where the trunk of the anterior tibial artery joins the arch. The convexity of this arch is towards the toes, and sends off the following branches.

(a,) The *first Digital Artery*, to the outer side of the little toe.

(b,) The *second Digital Artery*, which runs along the space between the two last metatarsal bones, and bifurcates into two branches, one to the inner side of the little toe, and the other to the outer side of the next toe.

(c,) The *third Digital Artery*, which bifurcates, in a similar manner, to the fourth and third toes.

(d,) The *fourth Digital Artery*, which bifurcates to the opposite sides of the third and second toes.

The concavity of the arch sends off *muscular* twigs, and also the *interosseous* arteries, three or four small twigs, which go to the deep-seated parts in the sole of the foot, and, perforating between the metatarsal bones, inosculate with the superior interosseous arteries on the upper side of the foot.

II. VEINS.

The Posterior Tibial Veins are generally two in number; they accompany the artery, and terminate in the popliteal vein; they are formed of branches, which correspond to those of the artery.

III. NERVES.

The Posterior Tibial Nerve, which is the continuation of the great sciatic nerve, sinks beneath the
soleus, and accompanies the posterior tibial artery down
the leg.   In the lower part of the ham, the nerve is still
posterior to the great vessels, but has passed obliquely
towards their inner side.   It gives off, in the ham, many
filaments to the muscles in its neighbourhood, and one
which traverses the opening of the interosseous ligament
to the fore part of the leg.   Where the posterior tibial
artery appears to bifurcate in sending off the peroneal
artery, the Nerve is seen lying between the two vessels ;
it soon attaches itself to the trunk of the posterior tibial,
and then descends on its outer or fibular side, adhering
intimately to the artery in the lower part of the limb, so
that where they pass along the sinuosity of the os calcis,
the nerve is situated close in contact with the side of
the artery, but nearer to the projection of the heel than
that vessel is.   It here gives off a filament to the integuments of the heel and foot, and, with the artery, divides, beneath the abductor pollicis, into two branches.

1. The *Internal Plantar Nerve,* proceeds directly
forwards covered by the abductor pollicis, as far as the
posterior extremity of the first metatarsal bone, giving
off some twigs : it then divides into *four branches,* which
supply the great toe, and the opposite sides of the three
next toes.

2. The *External Plantar Nerve* takes its course forwards and outwards, between the flexor digitorum brevis
and flexor accessorius, and, at the posterior extremity of
the fifth metatarsal bone, divides into (*a,*) *A superficial
branch,* which supplies the external border of the fifth
toe, and the opposed sides of the fourth and fifth toes.

(*b*.) *A deep-seated branch*, which penetrates to the adductor pollicis, interosseous, and transverse muscles.

SECTION III.

DISSECTION OF THE SOLE OF THE FOOT.

THE cuticle is very much thickened on the sole of the
foot, from constant pressure : betwixt the integuments
and plantar aponeurosis, we find a tough granulated fat,
intersected by numerous tendinous bands ; this fat adheres firmly both to the skin and to the aponeurosis, and
is dissected off with difficulty.

The PLANTAR FASCIA, or APONEUROSIS, is a very
strong tendinous expansion, which arises from the projecting extremity of the os calcis, and passes to the root
of the toes, covering and supporting the muscles of the
sole of the foot. It is of a triangular form : where it
arises from the heel, it is thick but narrow ; as it runs
over the foot, it becomes broader and thinner ; and it is
fixed to the head of each of the metatarsal bones by a
bifurcated extremity or process, which, by its splitting,
leaves room for the tendons, vessels, and nerves to pass.
It seems divided into three portions, which are connected
by strong fasciculi of tendinous fibres ; and fibres are sent
down, forming perpendicular partitions among the muscles, and separating them into three classes.

1. The middle portion, which is the largest and thick-
est, and under which are contained the flexor brevis
digitorum, and the tendons of the flexor longus and
lumbricales.

2. The external lateral portion, which is thin and
covers the muscle of the little toe.

3. The internal lateral portion, also thin, and concealing the muscles of the great toe.

On removing the plantar aponeurosis, it will be found that its inner surface adheres to the subjacent muscular fibres: the *first order* of muscles in the sole of the foot is now exposed, it consists of three muscles.

Abductor Pollicis, situated on the side of the great toe.

Abductor Minimi Digiti, on the side of the little toe.

Flexor Brevis Digitorum Pedis, the mass in the middle situated between the two abductors, extending from the os calcis to the toes.

I. ABDUCTOR POLLICIS PEDIS—*Arises*, tendinous and fleshy, from the lower and inner part of the os calcis; from the fore part of the same bone, from the internal annular ligament, and from the fascia plantaris: it stretches across the flexor tendons, and posterior vessels and nerve, where they pass from the leg into the foot; filling up the hollow between the projection of the heel and internal cuneiform bone.

Inserted, tendinous, into the internal sesamoid bone, and base of the first phalanx of the great toe.

Use: To move the great toe from the other toes.

2. ABDUCTOR MINIMI DIGITI PEDIS—*Arises*, tendinous and fleshy, from the inferior surface and outer side of the os calcis, and also from the fascia plantaris and septum, which divides it from the next muscle. Its fleshy mass lies on the outer border of the foot.

Inserted, tendinous, into the base of the metatarsal bone of the little toe, and into the outside of the base of the first phalanx.

This muscle can frequently be divided distinctly into two portions.

Use: To move the little toe from the other toes.

3. FLEXOR BREVIS DIGITORUM PEDIS PERFORATUS—*Arises*, fleshy, from the anterior and inferior part of the protuberance of the os calcis, and from the inner sur-

face of the fascia plantaris : also from the tendinous partitions betwixt it and the abductors of the great and little toe :—it forms a thick fleshy belly, which divides into four distinct portions, and sends off four tendons, which split for the passage of the tendons of the flexor longus digitorum, and are

Inserted into the sides of the second phalanx of the four lesser toes.

The tendon of the little toe is often wanting.

Use : To bend the second joint of the toes.

Situation : The muscles of this order are quite super-ficial, being only covered by the fascia plantaris.

The first order of muscles being removed, or being lifted from their origins, and left hanging by their ten-dons, the *second order* is exposed.

1. The tendon of the flexor longus digitorum pedis is seen coming from the inside of the os calcis; and, having reached the middle of the foot, dividing into its four tendons, which pass through the slits of the ten-dons of the flexor digitorum brevis, and are inserted into the base of the last phalanx of the four lesser toes. The flexor tendons, in their passage over the phalanges, are bound down by fibrous sheaths, and are invested by synovial membrane.

2. The tendon of the flexor longus pollicis is also seen passing forward from the hollow of the os calcis, and crossing under * the tendon of the flexor longus digitorum, and, having given to it a short slip of tendon, proceeding to the inner side of the foot; passing for-wards between the two portions of the flexor brevis pol-

---

* In the erect posture, it crosses above, lying nearer to the meta-tarsal bones than that tendon ; but, in the description, the sole of the foot is supposed to be placed uppermost.

licis, and between the two sesamoid bones, it is inserted into the base of the last phalanx of the great toe.

3. FLEXOR DIGITORUM ACCESSORIS, or *Massa Carnea Jacobi Sylvii—Arises*, fleshy, from the sinuosity at the inside of the os calcis, and, tendinous, from that bone more outwardly ; it forms a belly of a square form.

Inserted into the outside of the tendon of the flexor digitorum longus, just at its division.

Use : To assist the flexor longus.

4. LUMBRICALES PEDIS—*Arise*, by four tendinous and fleshy beginnings, from the tendons of the flexor longus digitorum, immediately after their division.

Inserted, by four slender tendons, into the inside of the first phalanx of the four lesser toes, and into the tendinous expansion that is sent from the extensors to cover the upper part of the toes.

Use : To promote the flexion of the toes, and to draw them inwards.

Situation : The muscles of the second order are covered and concealed by those of the first order : but the insertion of their tendinous extremities may be seen on removing the integuments.

The second order of muscles being removed, we expose the *third order.*

I. FLEXOR BREVIS POLLICIS PEDIS—*Arises*, tendinous, from the under and fore part of the os calcis, where it joins with the os cuboides : also from the os cuneiforme externum ; it forms a fleshy belly, which is connected inseparably to the abductor and adductor pollicis, and is divided into two portions, between which the tendon of the long flexor is situated.

Inserted, by two tendons, into the external and internal sesamoid bones ; and is continued on into the base of the first phalanx of the great toe. --

Use : To bend the first joint of the great toe.

2. ADDUCTOR POLLICIS PEDIS—*Arises*, tendinous and fleshy, from the under surface of the os cuboides, from the roots of the second, third, and fourth metatarsal bones, and from the ligaments connecting them; it forms a thick and short muscle, which is closely united with the external portion of the last muscle.

Inserted, tendinous, into the external sesamoid bone, and root of the metatarsal bone of the great toe. .

Use : To bring this toe nearer to the rest.

3. FLEXOR BREVIS MINIMI DIGITI PEDIS—*Arises*, tendinous and fleshy, from the os cuboides, from the root of the metatarsal bone of the little toe, and from the ligamentous sheath of the peroneus longus.

Inserted, tendinous, into the base of the first phalanx of the little toe, and into the anterior extremity of the metatarsal bone.

Use : To bend the toe.

4. TRANSVERSALIS PEDIS—Is a narrow muscle, scarcely an inch in breadth, extending across the foot, under the flexor tendons, close upon the heads of the metatarsal bones. It *arises*, tendinous, from the anterior or digital extremity of the metatarsal bone supporting the little toe; becoming fleshy, it crosses over the anterior extremities of the other metatarsal bones.

Inserted, tendinous, into the anterior extremity of the metatarsal bone of the great toe, and into the internal sesamoid bone, adhering to the adductor pollicis.

Use : To contract the foot, by bringing the toes nearer to each other.

Ranging with this order of muscles, we may also observe the termination of the tendon of the Tibialis Posticus, dividing into numerous tendinous slips, to be inserted into the bones of the tarsus.

2

Situation : The muscles of the third order lie under those of the second order, but are only partially concealed. The flexor brevis pollicis lies under and on each side of the tendon of the flexor longus pollicis : the adductor pollicis lies on the outer side of the flexor brevis, and is in part concealed by the tendons of the flexor digitorum longus.—The flexor brevis minimi digiti is a fleshy mass, lying on the metatarsal bone of the little toe, and not concealed by any muscle of the second order.—The transversalis pedis runs across under the tendons of the flexor digitorum longus and lumbricales, and is seen projecting betwixt those tendons.

Having removed the muscles last described, we expose the *fourth and last order.*

The termination of the tendon of the Peroneus longus is now exposed : it is seen descending from behind the outer ankle, and over the outer side of the os calcis, then passing along a groove in the os cuboides, and crossing the tarsal bones, to be inserted into the base of the metatarsal bone of the great toe, and into the internal cuneiform bone.

INTEROSSEI PEDIS INTERNI, are three in number, situated in the sole of the foot.—They *arise,* tendinous and fleshy, from between the metatarsal bones of the four lesser toes, and are

Inserted, tendinous, into the inside of the base of the first phalanx of each of the three lesser toes.

Use : To move the three lesser toes inwards towards the great toe.

INTEROSSEI PEDIS EXTERNI, are four in number, larger than the internal interossei, and situated on the back of the foot; they are *bicipites,* or arise by two slips, from the opposite surfaces of two metatarsal bones. They

Arise, tendinous and fleshy, between the metatarsal bones of all the toes.

Inserted, the first, into the inside of the base of the first phalanx of the second toe;—the second, into the outside of the same toe;—the third, into the outside of the middle toe;—the fourth into the outside of the fourth toe.

Use: To separate the toes.

CHAPTER VI.

DISSECTION OF THE HEAD.

SECTION I.

OF THE EXTERNAL PARTS OF THE HEAD.

THE integuments of the head are thick, and covered with hair; under the cutis there is a cellular substance which is much condensed, and which is closely connected with the epicranium, or expanded tendon of the occipito frontalis. This connexion renders the dissection of that muscle difficult. Make an incision from the root of the nose to the middle of the transverse ridge of the os occipitis, and reflect the skin on each side, taking care not to raise at the same time the tendinous expansion.

The Occipito-Frontalis is the only muscle which properly belongs to the hairy scalp; it is a single broad digastric muscle,

Arising, on each side of the head, fleshy and tendinous, from the transverse ridge of the occipital bone, as far forwards as the mastoid process: the fleshy fibres ascend, and form a broad thin tendon, which covers the whole upper part of the cranium ; this tendon terminates anteriorly in another expanded fleshy belly, which is

Inserted, on each side, into the skin of the forehead and eyebrows, intermixing with the fibres of the orbicularis palpebrarum and corrugator supercilii, and sending downwards, on each side of the nose, a slip of fibres, which spread out and cover the ossa nasi.

Situation : The posterior belly ranges with the upper borders of the two trapezii and sterno-mastoidei : the broad tendon, (sometimes termed the *Cranial Aponeurosis,)* adheres firmly to the skin, but very loosely to the pericranium, so as to admit of a sliding motion, and, laterally, is blended with the temporal fascia. The nasal slip is often described as a *Pyramidalis Nasi* on each side.

Use : To pull the skin of the head backwards, raise the eye-brows, and corrugate the skin of the forehead.

Many vessels and filaments of nerves ramify in the occipito-frontalis. The anterior arteries come out of the orbit, and consist of the superciliary and frontal branches of the ophthalmic artery ; posteriorly, we find branches of the occipital ; and, laterally, those of the temporal and posterior auris arteries. The anterior nerves also emerge from the orbit, and are filaments of the frontal nerves of the fifth pair ; the posterior fleshy expansion is penetrated by ramifications of the greater and less occipital nerves, derived from the cervical nerves ; and, finally, the lateral nerves consist of filaments of the portio dura, of the superficial temporal branch of the inferior maxillary, and of some ascending filaments of the cervical plexus.

The *Temporal Muscle,* covered by its aponeurosis, is observed on each side of the cranium ; but the description of this muscle belongs to the dissection of the face. Some small muscles of the external ear may be demonstrated at present.

1. *Muscles* moving the *External Ear.*

(1.) ATTOLLENS AUREM—*Arises* from the tendon of the occipito-frontalis, and from the aponeurosis of the temporal muscle. *Inserted* into the upper part of the root of the cartilage of the ear, opposite the antihelix. *Use :* To draw the ear upwards.

(2.) ANTERIOR AURIS—*Arises,* thin and membranous, from the posterior part of the zygomatic process of the temporal bone. *Inserted* into a small eminence on the back of the helix, opposite to the concha. *Use :* To draw the eminence a little forwards and upwards.

(3.) The RETRAHENTES AURIS—*Arise,* by two or three distinct slips, from the external and posterior part of the mastoid process, immediately above the insertion of the sterno-cleido-mastoideus. *Inserted* into the back part of the ear opposite to the septum, which separates the scapha and concha. *Use :* To draw the ear back, and stretch the concha.

Situation :—These muscles, placed between the skin and aponeuroses, do not always admit of clear demonstration. The Attollens aurem is a thin muscle on the side of the head, just above the ear : the anterior auris is a small muscle in front of the cartilage, and not always found : the retrahentes auris are two or three delicate slips, attaching the ear to the mastoid process.

2. The *proper muscles* of the *external ear* are very indistinct, and are only to be traced in a fresh muscular subject : they are placed on the surface of the cartilage, which gives form to the external ear. (1.) HELICIS MAJOR consists of some fibres, which cover the fore part of the helix, above the tragus. (2.) HELICIS MINOR

is smaller, and is placed behind and under the helicis major, on the projecting part of the helix, near the concha. (3.) TRAGICUS consists of a triangular fasciculus of fibres, covering the external surface of the tragus. (4.) ANTITRAGICUS is placed on the antitragus, occupying the space which separates it from the antihelix. (5.) TRANSVERSUS AURIS is found on the back part of the ear, arising from the prominent part of the concha, and extending to the outer side of the antihelix. *Use :* these muscles serve to produce the partial movements of these parts.

3. The *Muscles* of the *internal ear* are situated within the temporal bone itself: they are very small, and can only be seen when the internal parts of the organ of hearing are prepared. They consist of (1.) TENSOR TYMPANI, which *arises* from the cartilage of the Eustachian tube, and, reaching the tympanum, is *inserted* by a small tendon into the neck of the malleus. (2.) LAXATOR TYMPANI, is smaller, and passes from the spinous process of the sphenoid bone and side of the Eustachian tube, to be *inserted* into the long process of the malleus. (3.) STAPEDIUS is concealed within the cavity of the pyramid, from which its small tendon comes out, to be *inserted* into the neck of the stapes. The *Use* of these muscles is, by acting on the small bones of the internal ear, to modify the tension of the membranes with which they are connected.

SECTION II.

OF THE CONTENTS OF THE CRANIUM, OR THE BRAIN AND ITS MEMBRANES.

MAKE a transverse incision from ear to ear, over the crown of the head, through the tendon of the occipito-frontalis, and invert the two flaps on the face and neck. Remove the upper part of the cranium by the saw, which should be directed, anteriorly, through the frontal bone above the orbitar processes, and, posteriorly, as low as the transverse ridge of the occipital bone : the skull-cap or calvarium is then to be torn off, which requires considerable force from the adhesion of the subjacent dura mater. Remark, on the inner surface of the bone, the broad *sulcus* along the sagittal suture for the longitudinal sinus ; the numerous *grooves* corresponding with branches of the meningeal arteries ; the shallow *depressions* marking the convolutions of the cerebrum, and the small *fossæ* produced by the glandulæ Pacchioni.

The bone being removed, you expose the DURA MATER, a strong, dense, fibrous membrane, of a pearly-white or blueish appearance ; rough on its outer surface, and covered with bloody spots from the rupture of vessels, which connected it with the cranium. It is the outermost of the coverings of the brain, serving also as internal periosteum to the bones containing it ; it occupies the whole internal surface of the cranium, its adhesion being strongest in the line of the sutures, and at the base of the skull, where it sends prolongations round the nerves through the various foramina, and is continued with the external periosteum. It is also prolonged in a looser tubular form down the vertebral canal. It is described as separable into two laminæ, which form the triangular venous canals, named *Sinuses.* These sinuses return the blood from the brain, and will be noticed, as they are met with, in the course of the dissection.

The SUPERIOR LONGITUDINAL SINUS is to be examined at present ; it is situated in the middle convex line of the dura mater, extending from the crista galli, beneath the sagittal suture, to the middle of the os occipitis, where it bifurcates into the two *lateral sinuses.* Slit it open with the scissars ; observe its *triangular form ;* its *size,* enlarging as it proceeds backwards ; the numerous *openings of the veins* of the pia mater ; the *fræna,* or *chordæ Willisii,* strong slips of fibres crossing from side to side ; and the *Glandulæ Pacchioni internæ,* small, whitish, granular bodies, within the Sinus ; similar minute bodies are also found on the external surface of the dura mater, the *Glandulæ Pacchioni externæ.*

Observe the *arteries* of the dura mater ramifying on it, and projecting from its surface ; the chief artery is the *Arteria Meningea Media,* the great middle artery, which is a branch of the internal maxillary, and comes through the spinous hole of the sphenoid bone : it is seen ascending from the anterior inferior angle of the parietal bone, in a groove of which it lies, and bifurcating into its anterior and posterior branches : the dura mater has also anterior and posterior arteries, which are small and come off from other branches of the external carotid.

Divide the dura mater in the line of the division of the cranium, and invert it ; its internal surface is seen, smooth, glistening, and free from adhesion, except in the course of the longitudinal sinus, where the veins enter from the pia mater. Next proceed to examine the *Septa of the Brain,* or *Processes of the Dura Mater.*

1. The FALX MAJOR, SEPTUM CEREBRI, or *falciform process of the dura mater :*—to expose the falx, tear across the veins passing into the longitudinal sinus, and press to one side either of the hemispheres of the brain ; white granular bodies, similar to the *glandulæ Pacchioni,* present themselves in clusters, along the margin of the hemisphere : now examine this *falx major ;* it is a pro-

cess or duplicature of the dura mater, of an arched form, narrow anteriorly, but becoming broader posteriorly, containing in its upper convex border the superior longitudinal sinus, and passing down between the two hemispheres of the brain; it extends from the crista galli, along the middle line of the cranium, to the transverse ridge of the occipital bone, where it terminates in the middle of the next septum.

2. The TENTORIUM CEREBELLI, or *transverse septum*, is exposed partially by raising the posterior lobes of the cerebrum; this which is also termed the *tentorium cerebello super-extensum*, separates the posterior lobes of the cerebrum from the cerebellum, and will be more fully examined hereafter.

There is a third fold of the dura mater, not visible in this stage of the dissection. 3. The FALX MINOR, *falx of the cerebellum*, or *small occipital septum*, which will be seen when the cerebrum is removed, extending from the middle of the tentorium, between the two hemispheres or lobes of the cerebellum.

Detach the falx from the crista galli with your scissars, and turn it backwards to the occiput: observe in its lower concave edge the INFERIOR LONGITUDINAL SINUS, which is small, and enters a sinus in the tentorium, named the *fourth* or *straight* sinus. You have now exposed the *convolutions of the brain*, closely invested by the *tunica arachnoides* and *pia mater*. The TUNICA ARACHNOIDES is the more external of the two membranes, giving the smooth serous appearance; it is thin and transparent, and covers uniformly the surface of the pia mater, without passing into the interstices of its duplicatures: on the upper part of the brain, it is demonstrated with difficulty, being closely attached to the pia mater, but, by the blow-pipe, it may in general be raised into cells : on the base of the brain, it will be seen distinctly; and it is continued from thence into the vertebral canal around the spinal marrow. The Arachnoid is now considered to be a serous membrane, forming, like other membranes of that class, a shut sac, of which one portion envelopes the brain and spinal chord, with the pia mater interposed, while the outer portion or layer is reflected, at every point where nerves and vessels pass, to line the inner surface of the dura mater. It is also supposed to penetrate into the third ventricle, and from thence into the other ventricles, becoming continuous with the delicate lining membrane of those cavities.

The PIA MATER is the very vascular membrane, immediately investing the substance of the brain on all sides, and descending between its convolutions; it also passes between the plates or strata of the cerebellum, is continued over the spinal chord, and is prolonged around the nerves, forming their immediate investment or *neurilema*. Exteriorly, the pia mater is adherent to the arachnoid; its close connexion, interiorly, with the substance of the brain, is seen on tearing them apart, which produces the flocculent appearance, termed *Tomentum cerebri*. The pia mater enters the internal cavities of the brain, passing distinctly into the third and lateral ventricles, as will be seen in the course of the dissection.

The BRAIN is divided into *three parts*. 1. The *cerebrum*. 2. The *cerebellum*. 3. The *medulla oblongata*.

The CEREBRUM is the largest of these divisions, being the great mass filling the upper part of the cranium, and now exposed to view by the removal of the dura mater. Observe, on tearing off the pia mater, that the cerebrum is divided on its external surface into numerous undulated eminences or convolutions, *gyri*, with corresponding depressions, or *sulci*, between them, into which the pia mater enters. These sulci or clefts penetrate to the depth of an inch or more, and run in various directions.

On cutting off a slice from the brain, it is found to consist of two distinct substances : 1. The *Cineritious*, or *Cortical* substance, forming the outer part, of a softer consistence. 2. The *White*, or *medullary* substance, forming the inner part, of a fibrous structure, more dense, but varying in firmness of consistence, and having its surface dotted by red points from the division of blood vessels.

The Cerebrum is divided by a deep *Longitudinal fissure*, corresponding with the falx major, into two HE-MISPHERES ; and each hemisphere is subdivided into three LOBES.

1. The *Anterior lobes* rest on that part of the cranium, which forms the two orbits, and which is called the anterior fossæ of the base of the cranium.

2. The *Middle lobes* are situated before and above the medulla oblongata, and rest on the middle fossæ of the basis cranii, which are formed by the sphenoid and temporal bones.

3. The *Posterior lobes* are supported by the tentorium.

The division between the anterior and middle lobes is observable on the upper surface of the cerebrum : examine the anterior lateral part of the cerebral mass, near the temporal bone ; you discover a deep narrow sulcus, its sides united by vessels : this is the *Fissura Sylvii*, coming up from the base of the brain, and passing obliquely backwards between the anterior and middle lobes; it extends from the temporal ala of the os sphenoides to near the middle of the os parietale : the division between the middle and posterior lobes is not seen except on the base, where these two lobes are separated by a slight depression.

We now proceed with the dissection :—Gently separate with the fingers the two hemispheres ; you perceive an oblong, white, slightly convex body, the CORPUS CALLOSUM or COMMISSURA MAGNA, passing horizontally between them, and extending from before backwards ; it is about three inches in length, becoming somewhat broader posteriorly; laterally it is lapped over by the two hemispheres : it lies under the lower sharp edge of the falx, at the bottom of the longitudinal fissure, and incurvates downwards at both its extremities : on its surface is seen the *Raphe,* formed by *two longitudinal ridges,* with a superficial *furrow* between ; separate the ridges, and you perceive a *middle prominent line,* which is the proper *raphe,* penetrating through the substance of the corpus callosum :—the two *arteriæ callosæ* are seen arching backwards along the raphe, and, on each side, *medullary lines* or *striæ* pass *transversely* to it.

Slice away the two hemispheres horizontally within half an inch of the level of the corpus callosum : you observe, on each side, a central medullary portion, bounded by cineritious matter : these are termed *Centra ovalia lateralia* or *minora.* Continue your section, till it is on a level with the corpus callosum ; the large, oval, medullary appearance, thus produced, having the corpus callosum in its centre, is termed the MEDULLARY ARCH or CENTRUM OVALE : the continuity of the corpus callosum with the medulla of the brain is now displayed ; it unites the two hemispheres :—trace its *anterior extremity* passing forwards between the anterior lobes of the brain, and forming a *rounded medullary prominence* *,

* Observe the two *anterior cerebral* arteries curving round this

which curves downwards and then backwards, between the two corpora striata, and will be again observed at the base of the brain : the *posterior extremity* is also *rounded*, and incurvates downwards and then forwards, becoming continuous with the posterior crus of the *fornix*, on each side, and chiefly forming the *hippocampi* in the cornua of the lateral ventricle.

Under this central medullary arch are the two LA-TERAL VENTRICLES : perforate one of them on the side of the corpus callosum, half an inch from the raphe, and inflate gently with the blow-pipe; its extent will be seen, but if much force is used, the air will pass into the other cavity. The Ventricles should now be freely opened; run the knife cautiously along the two cavities, on each side of the raphe, and turn aside the divided medulla; you observe that they are separated from each other by a *medullary partition*, which descends from the inferior surface of the corpus callosum to the upper surface of the fornix; this is the SEPTUM LUCIDUM, and it appears partially transparent : this septum is triangular, with its apex behind, its base before ; it consists of two distinct medullary layers, with a narrow space between; pare away the corpus callosum, or cut transversely through it, reflecting the two portions gently from the septum ; the two layers of the latter will be made evident, with the intermediate cavity, which has been called the FIFTH VENTRICLE, or VENTRICLE OF THE SEPTUM LUCIDUM.

You now observe the extent and situation of the *Lateral Ventricles*, as they lie open on each side of the collapsed septum : they are triangular cavities of some

anterior prominence, to reach the upper surface of the corpus callosum, where each artery terminates in (1.) the *Arteria callosa*, running by the side of the raphe; and (2.) *another branch*, which ascends, in a tortuous form, on the inner surface of the hemisphere.

extent, situated in the middle of the cerebrum, lined with a fine membrane, and frequently containing some fluid : they begin in the anterior lobes, and proceed backwards parallel to one another; posteriorly they diverge : a vascular fold of pia-mater, the *plexus choroides*, is seen lying loose in each cavity. *Anteriorly*, each ventricle is bounded by the curved anterior portion of the corpus callosum ; the *roof* or *upper part* is formed by the under surface of the corpus callosum and adjacent medulla of the brain ; the *exterior wall*, by a greyish convex eminence, the *corpus striatum ;* the *septum lucidum* is the *inner wall* or partition ; and the *floor* is formed, outwardly, by an oval medullary body, the *thalamus nervi optici,*—inwardly, by the *fornix*, the *floating margin* of which is seen curving backwards along the ventricle, with the inner border of the choroid plexus dipping under it.

Each ventricle consists of a middle part or *body*, and three prolongations or *cornua*. Trace at present the *anterior cornu*, diverging outwards between the more acute convexity of the corpus striatum and the corpus callosum ; it is an angular cavity of small extent. The posterior and inferior cornua will be more readily traced in a future stage of the dissection, as they pass into the posterior and middle lobes of the brain.

Next examine the FORNIX; it is a flat medullary body, of a triangular shape, placed horizontally between the two lateral ventricles, and also separating them from the third ventricle which lies beneath. It is exposed on tearing away the septum lucidum, which is united to the upper surface; the lower surface is towards the third ventricle, resting on the optic thalami, but separated from them by the interposition of a vascular membrane, the *velum interpositum ;* the lateral margins of the fornix have been already pointed out in each ventricle, in the form of a flat and narrow medullary band ; this is termed the *corpus fimbriatum*, it is loose and unattached, curves round the inner side of the optic thalamus, becoming broader, and is continued along the ventricle into its inferior cornu.

Under the most anterior part of the fornix is the FORAMEN MONROIANUM : to discover this foramen, follow the course of the choroid plexus as it passes forwards in the ventricle, and gently turn the anterior part of the fornix to one side ; beneath it you perceive a slit, rather than a round hole, through which the plexus passes to the other side ; this is the *foramen of Monro ;* it is a space between the most anterior part of the convexity of the optic thalami and the anterior crus of the fornix ; the lateral ventricles communicate with each other, and with the third ventricle, by this opening, which is sometimes of considerable size.

The middle triangular part of the fornix is termed its *Body ;* the extremities are its *crura,* or *Pillars.* Divide transversely the body of the fornix, and invert it, by turning the anterior crus forwards, and the posterior crura backwards : on the under surface of the latter is an appearance of transverse lines, named *Corpus Psalloides, Psalterium,* or *Lyra**. The *anterior crus,* which appeared single, is now seen to split into two small medullary chords, which separate, and curve downwards before the optic thalami and between the two corpora striata. The two *posterior crura,* which are continued from the posterior angles of the fornix, pass backwards and diverge : Observe that the posterior crus, on each side, coalesces with the lower part of the corpus callosum,

---

* This appearance is frequently wanting.

and bends outwards, appearing to bifurcate; one, more slender, portion joins the medullary surface of the *hippocampus*, while the other, larger and flattened portion, adheres to the inner concave margin of the same hippocampus, forming the *Corpus Fimbriatum*, which has already been noticed, and which will again be seen extending along the inferior cornu of the ventricle.

The VELUM INTERPOSITUM is the vascular membrane, exposed by the inversion of the fornix, of triangular shape, and extending across from one plexus choroides to the other: it is stretched over the upper aperture of the *third ventricle*, covering the optic thalami and tubercula quadrigemina, and lining the inferior surfaces of the fornix and of the back part of the corpus callosum, under which it is continued from the exterior pia mater. Observe, in the middle of the velum, the *Vena Galeni*, consisting of two parallel branches, which run backwards, unite, and enter the *fourth sinus*. The velum is very broad posteriorly, and encloses in its folds the *Pineal Gland*.

The PLEXUS CHOROIDES is the granulated or vesicular mass, observed in the cavity of each ventricle; it is of a reddish colour, and is considered to be a vascular elongated fold of pia mater: on its inner side, each plexus is continuous, under the fornix, with the velum; anteriorly it is joined to its fellow through the foramen of Monro, and posteriorly it descends into the inferior cornu of the ventricle, its outer edge is loose and floating.

Detach the choroid plexus at its fore part, and turn it, with the velum, backwards: it will remain as a guide to the knife in tracing the inferior cornu of the ventricle.—You have now fully exposed

The CORPORA STRIATA, two pyriform cineritious convexities, situated in the fore part of each lateral ventricle, broad and prominent anteriorly, becoming narrow and diverging posteriorly, and continuous on their sides with the cerebral substance: they consist internally of the medullary and cortical substances disposed in striæ.

The THALAMI NERVORUM OPTICORUM are the two large, oval, medullary eminences, placed, by the side of each other, between the diverging extremities of the corpora striata ; towards their fore part is a peculiar eminence or convexity, called the *Anterior Tubercle ;* on their outer side, the thalami are continued with the corpora striata and substance of the brain : their inner sides form the lateral surfaces of the third ventricle; posteriorly they are unattached, being contiguous to the corpus fimbriatum ; below, they are united by the floor of the third ventricle.

The TÆNIA SEMICIRCULARIS GEMINÛM is the white medullary line, running in the angle or groove betwixt the corpus striatum and thalamus of each side.

Now trace the posterior cornu of the lateral ventricle, and the inferior cornu which descends into the middle lobe of the brain ;—follow the tract of the choroid plexus backwards from the body of the ventricle : it will lead first downwards and outwards, and then forwards and inwards :—where it takes the turn forwards, observe that the cavity bifurcates ; it is at this point that the *posterior cornu,* termed also the *Digital Cavity,* turns off, and stretches backwards into the posterior lobe of the brain ; it is about an inch in length, and contains a rounded medullary projection, the *Hippocampus minor.*

To follow the *inferior cornu,* which is more extensive, and appears the continued cavity of the ventricle, much of the lateral part of the cerebral substance must be sliced away ; the plexus choroides, which does not pass into the digital cavity, will conduct the knife ; lay open 'this inferior cornu, you perceive in it a large, white, medullary projection of the floor, the *Hippocampus Major*, or *Cornu Ammonis*; it is cineritious internally. Observe, that this larger prominence, (as well as the *hippocampus minor*,) is prolonged from the united posterior extremities of the fornix and corpus callosum; that it is at 'first narrow, is continued along the cornu, incurvating inwards, and that it terminates by a large *bulbous extremity*, with two or three projections separated by superficial grooves, which is named *Pes Hippocampi*: observe the thin floating edge, or band, on the inner concave side of the hippocampus, following the whole of its circuit, and curving round the optic thalamus; this is the *Tænia Hippocampi*, or *corpus fimbriatum* of the fornix, before seen in the body of the ventricle : and between the optic thalamus and tænia is the chink or fissure, by which the pia mater enters the lateral ventricle from the base of the brain. Observe, that the outer margin of the hippocampus is bounded by a narrow sulcus, or depressed line, exterior to which is another *longitudinal medullary eminence*, following the outer border, and of much the same appearance : this *Pes accessorius*, or *collateral eminence*, is sometimes as distinct as the hippocampus itself. Lift up the tænia hippocampi, and you perceive, beneath it, a *narrow, brownish-coloured band*, minutely *indented* on its surface *, and continued, at its posterior extremity, with the inner part of the corpus callosum.

At this stage of the dissection, you may trace the continuity of the velum interpositum with. the external pia mater : divide carefully the posterior border of the corpus callosum, and raise it on either side of the inci-

* *Fascia Dentata.*

sion, and the continuation of membrane, beneath it, is evident. There is here a manifest communication between the exterior surface of the brain, and its internal cavities, by a large *transverse fissure*, which is placed beneath the posterior border of the corpus callosum, and which is continued, on each side, with the *semicircular fissures* observed between the corpus fimbriatum and thalamus *.

We now proceed to the examination of the *third Ventricle*, and of the *Pineal Gland*. Separate gently the two optic thalami; you perceive an exceedingly soft, broad, cineritious substance, uniting their flattened internal surfaces; this is the *Commissura Mollis*, and it is immediately broken through :—The narrow, longitudinal, fissure now exposed, is the THIRD VENTRICLE; it is closed, superiorly, by the *velum* and *fornix*; its sides are formed by the two *optic thalami*; inferiorly, its floor consists of a thin *medullary plate*, which unites the thalami below, and which is made up by certain parts to be seen at the base of the brain; anteriorly, it is bounded by the *commissura anterior* and bifurcated *anterior crus* of the fornix; and, posteriorly, by the *commissura posterior* and *tubercula quadrigemina*. Observe the *anterior commissure*, at the fore part of the ventricle, a short, cylindrical, medullary chord, stretched transversely between the lower anterior parts of the corpora striata †, and immediately in front of the diverging peduncles of the anterior crus of the fornix; the

---

* These three fissures, thus united, are the *Great Cerebral Fissure* of Bichat.

† Cut the corpus striatum horizontally on a level with the anterior commissure, and this medullary chord will be seen proceeding deeply into the substance of the hemisphere, spreading out and curving backwards.

slit, or interval, here formed by the crus bifurcating, is named *vulva*, or *foramen commune anterius*; it is the space by which the three ventricles communicate, and is a prolongation downwards of the foramen of Monro, being bounded posteriorly by the optic thalami. Beneath the anterior commissure the third ventricle appears of greater depth, and dips downwards; this is the opening of the passage, named the *Iter ad Infundibulum*: it leads to a short funnel, or conical process, of grey substance, which terminates in the pituitary gland: observe the anterior peduncles of the fornix, arching downwards, behind the iter ad infundibulum, to terminate in the *corpora albicantia* or *mamillaria* at the base of the brain.

At the back part of the third ventricle, observe the *Commissura Posterior*, shorter and thicker than the anterior, crossing transversely in front of the tubercula quadrigemina; immediately below this commissure is the opening of a canal, which extends, beneath the corpora quadrigemina and pineal gland, from the third into the fourth ventricle: this is named the *Aqueduct of Sylvius* or *Canalis Medius*:—the *Anus*, or *foramen commune posterius*, is the chink or opening formed on first separating the optic thalami, between the commissura mollis and posterior commissure, and it leads into the back part of the third ventricle; but this opening is closed by the velum stretched across, until that membrane is removed.

Continuing to reflect backwards the velum, you discover, involved in its layers, the PINEAL GLAND, or *Conarium*, a soft, reddish-gray, conical body, of the size of a pea, placed immediately behind the commissura posterior, in the depression between the uppermost pair of the tubercula quadrigemina; it is isolated from the cerebral substance, except at its broader anterior part, from which two white conical processes, or small *Peduncles*, pass to the optic thalami. Squeeze the gland between the fingers; it generally contains some gritty matter, resembling sand, of a pale yellowish colour, and named by Soëmmering *Acervulus Glandulæ Pinealis*.

Detach the pineal gland with the folds of velum :— cut away the posterior lobes of the cerebrum, and slit the tentorium on each side. Then press back the projecting middle part of the cerebellum *, and observe

The TUBERCULA QUADRIGEMINA, four small white rounded eminences, lying under the pineal gland, behind the third ventricle and above the fourth, adhering together, but separated externally by a cruciform groove: the upper and anterior are the largest, and are named the *Nates;* the posterior are called *Testes :* the nates are immediately behind the posterior commissure.

From the under part of the testes, there projects backwards, a thin greyish medullary lamina, which is united laterally to two rounded white chords or pillars; this is the *Valve of Vieussens,* or *Valvula Cerebri,* and its lateral pillars, ascending from the cerebellum, are termed the *Processus à Cerebello ad testes,* or *Superior Peduncles of the Cerebellum.* This valve is the roof of the fourth ventricle, as will be seen on introducing a probe or blow-pipe from the third ventricle into the canalis medius : below, it is united to the posterior wall of the fourth ventricle. On the exterior surface of the

---

* The Arachnoid may be observed coming from the surface of the cerebellum, and passing under the velum ; and it is here that it is said to penetrate into the third ventricle, by an oval canal, which, having been first described by Bichat, is termed the *arachnoid canal* of Bichat.

valve, close to the base of the testes, observe the *origin* of the *nervus patheticus* or *fourth pair*, which nerve is seen, thread-like *, winding round under the edge of the tentorium.

Remark the TENTORIUM, stretched over the cerebellum. It is formed by the laminæ of dura mater, which are reflected off from the os occipitis and temporal bones, to be fixed anteriorly to the clinoid processes of the sphenoid bone; on its fore part, it has a great notch, or *Oval Foramen*, through which the crura cerebri pass. Slit up the tentorium, and observe the FALX MINOR, or *falx of the cerebellum*, coming off from its centre, opposite to the *falx major*, and passing downwards between the hemispheres of the cerebellum, to the posterior margin of the foramen magnum. Trace the bifurcation of the *longitudinal* into the two *lateral sinuses*; these sinuses are formed by the splitting of the laminæ of the tentorium, and follow the course of that septum, occupying the lateral grooves of the occipital bone, and then dip downwards, on each side, through the foramen lacerum in basi cranii, to terminate in the internal jugular veins; the right lateral sinus is generally the largest. Observe the *fourth* or *straight sinus*, running along the middle of the tentorium, and joining the longitudinal sinus at the point where it bifurcates; and entering the same point, the two *occipital sinuses*, which come up in the falx cerebelli: the *Venous Cavity*, which is thus formed by the junction of six sinuses, at this point of union of the three great folds of dura mater, is named *Torcular Herophili*.

To examine the *Cerebellum in situ*, it will be conve-

---

* The roots of this nerve, varying from one to four, are very soft and easily broken.

nient to remove with the saw the posterior part of the os occipitis.

CEREBELLUM. This division of the brain is situated in the fossæ of the os occipitis, under the posterior lobes of the cerebrum, and is divided into two hemispheres or lobes by a fissure, which receives the *falx minor*. *Externally*, it has a *laminated* appearance, consisting of numerous flat strata of cineritious substance, disposed in a semicircular manner, and separated by narrow but deep *sulci*, into which the pia mater enters, while the arachnoid passes over them. *Internally*, it is formed of the medullary substance.

The upper surface of the cerebellum exhibits its two lateral flattened hemispheres, with a prominent central part, the *Superior Vermiform Process*. Trace the vertical fissure separating the hemispheres, and observe at its anterior inferior part, behind the medulla oblongata, the *Inferior Vermiform Process*, a convex eminence, connecting the two hemispheres below, which will be again observed on the base of the brain.

We now proceed to examine the FOURTH VENTRICLE, or *Ventricle of the cerebellum*. Introduce a probe, as before directed, along the *canalis medius;* it will be seen to raise the *Valve of Vieussens:* divide this medullary layer longitudinally, and the cavity of the fourth ventricle will be laid open : or it may be exposed by a deep perpendicular incision through the cerebellum. It is a small triangular cavity, with its base upwards, and inclined obliquely downwards and backwards; its *anterior wall* is formed by the posterior surface of the pons Varolii; *laterally*, it is bounded by the processus à cerebello ad testes; *posteriorly* by the cerebellum; *above*, it is covered by the *valve of Vieussens*, and it is closed *below* by the pia mater only, which is, however, firm

and resisting. Observe, on its anterior wall a *groove*
or *fissure*, which, terminating below in a sharp point, is
named *Calamis Scriptorius* : on each side of the groove
are seen several *transverse medullary lines,* which are
the origin of the auditory nerve. The *Canalis Medius*
enters the upper part of the fourth ventricle, and a *small
plexus* is observed in its lower part, termed the *choroid
plexus of the fourth Ventricle* *.

Make a vertical section through either hemisphere of
the cerebellum : the arborescent appearance of the medullary substance thus produced is termed the *Arbor
Vitæ;* and the trunks of medullary matter proceed from
each hemisphere to the pons Varolii, and are named the
Peduncles of the Cerebellum, or *Processus ad pontem:*—
the *processus ad testes* were before demonstrated, and
are seen ascending on the side of the valve of Vieussens.
Lift up the hemispheres of the cerebellum, and you perceive a *third pair of peduncles,* which pass from the
lower part of the cerebellum to the posterior surface of
the medulla oblongata and spinal chord : these are
called the *Processus ad medullam spinalem,* and, in the
dissection of the base of the brain, they will again present themselves as the *corpora restiformia.*

The BASE OF THE BRAIN is now to be demonstrated,
for which purpose the remaining portion of the mass is
to be removed from the cranium, and examined in an
inverted position. Raise it carefully from the cavity of
the skull, beginning anteriorly ; separate the olfactory
bulbs from the fossæ on the sides of the crista galli ;
cut across the optic nerves, and carotid arteries, with
the delicate infundibulum, at the sella turcica; and,
continuing to turn back the mass, divide the other ce-

---

* *Plexus Halleri.*

rebral nerves, on each side, close to the foramina by which they leave the skull: the knife is then to be pass-ed deep into the vertebral canal, and the spinal chord, with the vertebral arteries, cut across. This will enable us to remove the brain, and to place it on the table, with the base upwards, for demonstration.

First observe the general appearance of the inverted mass: anteriorly, and on the sides, the cineritious infe-rior surface of the *anterior*, and *middle lobes;* posteri-orly, the *cerebellum;* and, in the central part, uniting the cerebrum with the cerebellum, the medullary pro-minence of the *Pons Vorolii.* The *tunica Arachnoides,* which was demonstrated with difficulty on the upper surface of the brain, is here very evident, appearing loose, and unattached to the subjacent pia mater, and stretching from the optic nerves backwards to the Pons.

The chief trunks of the *arteries of the brain* are seen on its base, uniting, and forming the *circulus arterio-sus of Willis.* The INTERNAL CAROTID ARTERY emerges from the cavernous sinus at the side of the sella turcica : its divided trunk will be found on the base of the brain, by the side of the optic commissure, giving off its three branches. It first sends backwards the *Ramus com-municans,* a small artery of the length of an inch, which runs, covered by the arachnoid, along the side of the infundibulum and corpus albicans, to join the posterior cerebral branch of the Basilic : the carotid then termi-nates in two branches; (1.) The *A. cerebri anterior,* enters the fissure between the anterior lobes of the brain, and will be found to communicate with its fellow, by a large, but short, *transverse branch,* which completes the circle of Willis at the fore part. It is then continued forward over the anterior extremity of the corpus callo-sum. (2.) The *A. media cerebri,* dips into the fissure of Sylvius, and is distributed to the anterior and middle lobes of the brain.

The two VERTEBRAL ARTERIES are seen ascending on the sides of the medulla oblongata ; then uniting at the posterior edge of the pons Varolii, to form the BASILAR ARTERY, which is lodged in the middle groove of the pons, covered by the arachnoid membrane. At the anterior margin of the pons, the Basilar terminates in four branches, two on each side : (1.) The *Superior artery of the cerebellum*, which passes outwards, round the pons Varolii, to the upper surface of the cerebellum. (2.) The *A. Cerebri posterior*, which receives the ramus communicans of the carotid, and passes to the posterior lobe of the brain. The nerve of the *third pair* is observed coming out between these two branches of the basilar artery.

The *inferior surface* of the cerebral *lobes* may now be examined. First observe the *anterior lobes*, with the inferior termination of the *longitudinal fissure :* on each side is seen the OLFACTORY NERVE, or *first pair*, a flat pulpy chord, lodged in a superficial groove of the anterior lobe, running exterior to the pia mater, but covered and tied down by the arachnoid, and terminating anteriorly in a small *oval ganglion*, or *Bulb*, of a greyish colour and very soft. Next remark the *middle lobes*, each of which has a projecting fore part or *Monticulus*, and is separated from the anterior lobe by a transverse fissure, the *Fissura Sylvii :* the two surfaces of this fissure are closely united by a web of vessels, on dividing which the *middle cerebral artery* is seen deeply seated in the fissure. The two *anterior lobes*, are also closely connected to each other by the arachnoid passing across, and by *vascular telæ*, or *webs ;* tear asunder this membranous connexion, and, on separating the two lobes, you expose the two *anterior cerebral arteries*, running parallel to one another : you also discover the *anterior extremity* of the *corpus callosum*, which is observed to be continued backwards, embracing laterally the anterior parts of the *corpora striata*, and ending in two *medullary striæ*, which pass, on each side, towards the fissura Sylvii :—between this part of the corpus callosum, and the upper * surface of the optic commissure, is a *greyish pulpy lamina*, which closes the lower anterior part of the third ventricle. Trace the olfactory nerve backwards into the fissura Sylvii ; it will be seen sinking between the anterior and middle lobes of the brain, and spreading out in a *triangular form :* it will be found to arise by three roots, of which the external passes outwards, and may be traced to the under part of the corpus striatum ; the other roots spring from the medulla of the anterior lobe; the olfactory bulbs, or *processus mamillares*, occupy the cribriform plates on each side of the crista galli, and send numerous filaments through the foramina to the mucous membrane of the nose.

Observe that the fissura Sylvii is continued, at its posterior and inner extremity, into a *longitudinal fissure*, which is bounded outwardly by the middle lobe, within by the crura cerebri ; it is by this fissure, on each side, that the pia mater enters the lateral ventricle : the medullary surface, at the point of union of these two fissures, is remarkably *perforated* by numerous small apertures for vessels.

Next observe the OPTIC NERVES, or *Second Pair*, which were cut across, and removed from the foramina optica ; trace the two nerves backwards, converging to their *union* or *commissure* †, which was placed on the

---

* Inferior in this inverted position of the brain.

† *Chiasma nervorum opticorum.* Opinions are much divided as to

fore part of the sella turcica. Strip off the pia mater, and examine the *tractus nervi optici*; it is seen, on each side, passing backwards from the commissure, in the form of a *flattened medullary band*, which winds circularly over the root of the crus cerebri, adhering to the tuber cinereum, and, becoming broader, is lost in the back part of the optic thalamus and tubercula quadrigemina. Observe a small *greyish tubercle* on the outer margin of the tractus, just where it is disappearing, and another similar nodule, on its inner side, close to the posterior margin of the crus cerebri; these are the *corpora geniculata*.

Immediately behind the optic commissure is the *Tuber Cinereum*, a grey quadrilateral substance, united anteriorly to these nerves, and posteriorly to the corpora albicantia and crura cerebri; it assists to form the floor of the third ventricle. In the middle of this grey substance, we discover a reddish-coloured body, or process, of a conical figure, the *Infundibulum*, which, in removing the brain, was divided from its connexion with the pituitary gland. Turn to the base of the skull, and examine the PITUITARY GLAND; it is a globular, soft, reddish body, lodged in the sella Turcica, or pituitary fossa of the sphenoid bone, surrounded on all sides by the dura mater, except its upper surface, which is covered by the arachnoid*; it is evidently composed of two portions; the anterior, kidney-shaped, large and more firm; the posterior, rounded and soft:—the infundibulum terminates in the upper surface of the gland, but will not be found pervious.

the partial or complete decussation, or simple juxta-position of the fibres of the two nerves.

* The pituitary gland is surrounded by the *Circular* or *Coronary Sinus*, which opens on each side into the *Cavernous Sinus*.

Behind the infundibulum and grey substance, we observe two medullary, pea-like, bodies, the *Corpora Albicantia, Mamillaria,* or *Pisiformia;* they are united to each other, by a *minute whitish chord;* the anterior peduncles of the fornix terminate in these globular bodies, and they form a part of the floor of the third ventricle. Behind these bodies, and between the crura cerebri, we remark a deep *triangular pit* or *excavation,* occupied by medullary substance, which also forms a part of the floor of the third ventricle, and is pierced by numerous vessels: this is by some named the *Pons Tarini;* and here the two *Nerves* of the *third pair* take their origin.

Now examine the PONS VAROLII, or TUBER ANNU-LARE *; dissect off the pia mater, which adheres closely to its surface. It is the *rounded medullary eminence,* observed in the middle of the base of the brain, uniting the cerebrum and cerebellum; in this inverted position, it is stretched over the *crura* or processes of these two organs, like a bridge: the *crura* or *Peduncles* of the *Cerebellum* are two medullary trunks proceeding from that part of the cerebellum, which was seen to form the arbor vitæ, and uniting at the pons: the *Crura Cerebri* are also two large medullary pillars, fasciculated on their surface, coming from the middle part of each hemisphere of the brain, more immediately from the corpora striata: they are seen emerging from beneath the tractus opticus, on each side, uniting at an acute angle, and also terminating in the Pons, passing under the crura of the cerebellum.

On the surface of the Pons Varolii is seen a broad *longitudinal sulcus,* with *transverse striæ* passing from it on each side: this sulcus lodges the Basilar artery;

---

* *Nodus Encephali.*

it terminates, anteriorly, by a circular depression between the crura cerebri, termed the *foramen cæcum anterius,* and posteriorly, at the point of union between the pons and corpora pyramidalia, by the *foramen cæcum poste-rius.* In the natural position, the Pons Varolii lies on the basilar process of the occipital bone. Make a trans-verse section of either crus cerebri near the pons, and you discover a *dark patch* of a brown colour; this is named *Locus Niger,* or *Substantia Nigra.*

The MEDULLA OBLONGATA is that portion of the brain, which extends from the pons varolii downwards to the foramen magnum ; it is separated from the pons by a deep groove, and is of a pyriform shape ; its surface is raised- into four eminences. (1.) The two *internal* medullary prominences are termed the *Corpora Pyra-midalia :* they are separated by a longitudinal furrow, in which four or five transverse medullary chords are observed. (2.) The two *external* elevations are named *Corpora Olivaria,* a section of these exposes an irregular cineritious substance, the *corpus fimbriatum olivæ.* On the outer sides of the corpora olivaria, and below them, we perceive the *Corpora Restiformia* *, termed also *Pyramidalia lateralia,* which have already been noticed as the processes sent off from the cerebellum to the upper part of the spinal marrow. The medulla oblongata now contracts itself, and passing through the foramen mag-num occipitale, assumes the name of *Medulla Spinalis* or the *Spinal Chord.*

The two first pairs of Nerves have already been de-

---

* Mr. C. Bell's *respiratory tractus* or *column* is a narrow chord of medullary matter, situated between the corpus olivare and restiforme, and traced down the spinal marrow, between the anterior and poste-rior roots of the spinal nerves. It gives origin to his respiratory class of nerves.

scribed : we now proceed to examine the origin, and course within the skull, of the remaining *cerebral Nerves*.

The THIRD PAIR, or *Nervi motores oculorum*, arise from the inner part of the crura cerebri, and have been already noticed, coming out from the pit or hollow between the two crura. The nerve of each side passes forwards and outwards, and perforating the dura mater on the outer side of the posterior clinoid process, is lodged in a canal, which is formed in the outer wall of the cavernous sinus ; it then passes through the foramen lacerum orbitale superius to the muscles of the eye.

The FOURTH PAIR, or *Nervi Pathetici*, which are extremely slender, and the smallest of the cerebral nerves, have been already seen arising from the posterior surface of the valve of Vieussens, near the testes, and passing round immediately under the edge of the tentorium : this nerve comes out, thread-like, from between the cerebrum and cerebellum, advances forwards over the crura cerebri, and, piercing the dura mater behind the posterior clinoid process, passes through the cavernous sinus, and enters the orbit by the foramen lacerum superius, to terminate in the superior oblique muscle of the eye : it is separated from the cavity of the sinus by a very fine and cellular membrane.

The FIFTH PAIR, or *Trigemini*, are the largest of the cerebral nerves. The nerve, on each side, arises by separate filaments from the anterior part of the crus cerebelli, where the crus unites with the pons Varolii : these filaments are very numerous, and form a small anterior and internal fasciculus, and another broad and external ; the two fasciculi pass together, forming a pretty large trunk, obliquely outwards to the upper edge of the petrous portion of the temporal bone, and penetrate the dura mater ; here, close on the outside of the cavernous sinus, and beneath the dura mater, you find‑ the larger fasciculus untwisting itself, and forming a flat‑ irregular ganglion, the *Ganglion Gasserianum;* from the anterior convex border of this ganglion pass off three great branches: (1.) The *Opthalmic nerve,* or *ocular branch,* which enters a passage in the outer wall of the cavernous sinus, and passes through the foramen lacerum superius into the orbit. (2.) The *superior Maxillary Nerve* passes through the foramen rotundum to the upper jaw and face. (3.) The *inferior Maxillary Nerve* is joined by the narrow anterior fasciculus, (which passes under the ganglion without intermixing with it,) and runs through the foramen ovale to the lower jaw and tongue.

The SIXTH PAIR, *Motores oculorum externi,* or *Abductores.* This nerve is small, but not so slender as the fourth pair; it is observed arising, at a short interval from its fellow, from the corpora pyramidalia, and posterior margin of the pons Vorolii, and then passing forwards over the surface of the pons, to which it is pretty closely connected. It perforates the dura mater at some distance below the posterior clinoid process, and enters the cavernous sinus *; there it runs on the outer side of

---

* The *Cavernous Sinuses* are formed between the two laminæ of the dura mater, on the sides of the sella turcica. Each of these sinuses contains, within its boundaries, the internal carotid artery, the third and fourth pair of nerves, the first branch of the fifth pair, and the sixth pair. It should be slit open and examined; the carotid and nerve of the sixth pair lie apparently bathed in the blood of the sinus, but are separated from the fluid by a reflection of the lining membrane. The interior of the sinus is crossed by a number of reddish filaments, giving it a cellular appearance; these appear to consist of reflections of the internal membrane, and of nervous filaments, two of which are united to the nerve of the sixth pair; a minute ganglion, termed by Cloquet and others the *cavernous ganglion,* is sometimes observed on

the internal carotid, being in contact with the artery, and receiving generally two filaments from the superior cervical ganglion of the great sympathetic; the nerve then passes through the foramen lacerum orbitale superius, to the rectus externus muscle.

The SEVENTH PAIR. This nerve is observed, on each side, behind the root of the crus cerebelli, more outwardly than the sixth pair; it consists of two portions, which are closely connected. (1.) The *Portio Dura*, or *Facial Nerve*, arises from the posterior border of the pons Varolii, near its union with the corpus restiforme, a few lines more outwardly than the sixth pair. (2.) The *Portio Mollis*, or *Auditory Nerve*, arises immediately behind the portio dura, from the surface of the corpus restiforme: the root of the nerve is covered by a small layer of grey matter, and it may be traced to the white striæ on the inner surface of the fourth ventricle. Both portions of the seventh pair come out of the triangular fossa between the corpus olivare, crus cerebelli, and pons Varolii, the facial nerve being received into a groove on the surface of the portio mollis. The two nerves proceed together, accompanied by an artery, to the meatus auditorius internus, where the portio mollis is distributed to the parts of the internal ear, while the portio dura runs through the aqueduct of Fallopius, and emerges from the stylo-mastoid foramen to form the principal nerve of the face.

The EIGHTH PAIR, or Par Vagum, is a broad flat nerve, observed immediately behind the seventh pair; it arises by numerous separate filaments from the groove

the outer side of the carotid; these nervous filaments are continued from a delicate *plexus*, which is found surrounding the internal carotid in its canal, and which is formed by ascending twigs from the superior cervical ganglion of the great sympathetic.

between the corpus olivare and corpus restiforme : two distinct fasciculi are formed ; the uppermost, called *Nervus glosso-pharyngeus*, is composed of the first four or five filaments ; the remainder unite and form a flattened band, which is the proper *Nervus Vagus*, or *Pneumogastric nerve*. These two divisions of the eighth pair, (now usually considered as distinct nerves,) run together towards the foramen lacerum in basi cranii, and pass out through the anterior part of this hole, in separate sheaths of dura mater,-having been first joined by a *third nerve*, the *Nervus accessorius ad par vagum*, which is seen running up from the lateral part of the medulla spinalis through the great occipital foramen : this accessory nerve again leaves the skull in the same sheath with the nervus vagus. The great *Lateral sinus* passes out by the back part of the same foramen, to form the internal jugular vein ; it is separated from the nerve by a slip of cartilage.

The NINTH PAIR, *Linguales* or *Hypoglossi*. This nerve arises from the furrow between the corpus oblivare and corpus pyramidale, by several distinct filaments, which often pierce the dura mater separately. It passes through the anterior cóndyloid hole of the occipital bone, to supply the muscles of the tongue.

The TENTH PAIR, or *Suboccipitales*, arise on each side, by two bundles, from the extremity of the medulla oblongata, and upper part of the spinal marrow ; pass through the dura mater by the same foramen which gives entrance to the vertebral artery, and run through the foramen magnum occipitale to the muscles at the base of the cranium. These are now frequently described as the first pair of cervical nerves.

The inferior surface of the cerebellum remains to be examined, which will conclude the demonstration of the brain. First, observe a *deep depression* between the two hemispheres, in which the medulla oblongata is lodged: on each side the cerebellum swells into an *oval convex eminence*, and, anterior to this, an oblong slightly prominent surface, lying under the nerve of the eighth pair, has been called the *lobule of the nervus vagus*. Lift up the spinal chord, and you find these lateral convexities of the cerebellum continued below into the *Inferior Vermiform Process*. By raising the chord, you have now opened the lower part of the *fourth ventricle;* the red granulated mass named its *choroid plexus,* is seen bifurcating into two lateral portions, which pass out of the cavity *. Observe that the ventricle is bounded posteriorly by two *globular cineritious eminences,* and immediately before the vermiform process, there is a *projecting tubercle,* composed of transverse lamellæ, and united by its base to the cerebellum.

Cut into the medullary trunk of either hemisphere of the cerebellum, parallel to the medulla oblongata, at the distance of half an inch; you display an *ovoid central portion* or *nucleus*, which is separated from the surrounding medulla by an *indented cineritious line :* this is named the *corpus dentatum* or *rhomboideum* of the cerebellum.

* M. Magendie contends, that there is a natural opening at the lower part of the calamus scriptorius, which permits a constant communication of the *cephalo-spinal fluid* with the internal cavities of the brain. This fluid is said to exist constantly during life, between the pia mater investing the brain and spinal marrow, and the layer or reflection of arachnoid immediately covering it; but it disappears in some hours after death.

SECTION III.

OF THE VESSELS OF THE BRAIN.

THE ARTERIES ramify largely on the pia mater, before they enter the substance of the brain; they come from two large branches.

1. The INTERNAL CAROTID ARTERY enters the skull, by a winding course through the carotid canal of the temporal bone. After passing through the cavernous sinus, and perforating the dura mater beneath the anterior clinoid process, it divides into three branches : 1. The *Anterior cerebral artery*, advancing forwards between the two anterior lobes. 2. The *Middle cerebral artery*, entering the fissura Sylvii. 3. The *Communicating Artery*, uniting with the basilar. The carotid, while emerging from the cavernous sinus, gives off the *Ophthalmic Artery*, a considerable vessel, which passes through the foramen opticum to the eye and its appendages.

2. The VERTEBRAL ARTERY arises from the subclavian, ascends through the foramina in the transverse processes of the cervical vertebræ, turns horizontally along the atlas, and, then penetrating the dura mater, enters the cranium by the foramen magnum ; it ascends between the basilar process of the occipital bone and lateral part of the medulla oblongata, giving off some twigs to the dura mater, an *anterior* and *posterior spinal artery* to the spinal marrow, and the *Arteria cerebelli inferior vel posterior* to the inferior surface of the cerebellum : then uniting with its fellow, it forms the BASILAR ARTERY, which terminates in the two *Superior arteries of the cerebellum,* and two *Posterior cerebral arteries.*

K

The distribution and anastomoses of these *cerebral.
arteries* have been observed in the dissection of the base
of the brain.

The Veins pour their blood into the sinuses, of which
the principal have been described in the dissection of
the brain; they may be recapitulated as the *superior
longitudinal,* the two *lateral,* the *inferior longitudinal,*
the *straight sinus,* two *occipital,* and two *cavernous
sinuses,* and the *circular sinus.* Some other sinuses are
also found at the base of the brain, and are generally
enumerated, as two *superior petrous,* two *inferior petrous,* and a *transverse sinus.* They all terminate in
the lateral sinuses, and internal jugular veins; these
sinuses are lined by a fine membrane prolonged from the
venous system.

———

SECTION IV.

OF THE SPINAL CHORD AND ITS NERVES.

This part of the nervous or sensorial system will be here
described, although its dissection cannot be performed
till all the muscles of the back are removed, so that the
posterior part of the spinal canal may be sawed off.

On opening the *Spinal Canal,* we perceive the *Spinal
Chord,* or *Marrow,* enveloped by the same membranes
as the brain. The dura mater forms a complete sheath,
the *Theca Vertebralis,* which invests the chord through
the whole canal of the vertebral column; this sheath is
smaller than the vertebral canal, and is connected with
it closely only at the anterior part. On opening this
sheath, we see the *Medullary Chord* lying loose within
it, and invested by the pia mater * and tunica arachnoides.

* The pia mater of the chord differs considerably in appearance
from the same membrane within the skull; it is of denser texture,
and few vessels ramify on it.

Divide transversely the Spinal Marrow; you observe
that it is formed externally of medullary substance, internally of cineritious, which latter consists of a middle
transverse, and two lateral curved portions. The chord
runs down, varying in thickness at different parts, to the
first lumbar vertebra, where it terminates by numerous
nervous fasciculi, which, continuing to descend in the
sacral canal, form the *Cauda Equina.* It is closely
embraced by the pia mater, to which membrane the
arachnoid is attached very loosely. Observe, on each
side, the *Ligamentum Denticulatum,* a membranous
connexion between the dura and pia mater, running
down the chord between the anterior and posterior roots
of the spinal nerves, adhering by its inner border to the
pia mater, its outer border attached to the dura mater
by distinct pointed slips, or processes. The arteries of
the spinal chord are also seen running down on its anterior and posterior surfaces; the *posterior,* as two separate trunks, with frequent transverse anastomoses; the
anterior, united below the foramen magnum into a single
vessel, which descends in the median line. The veins
or *Vertebral Sinuses,* consist of two long trunks, which
descend, laterally, on the anterior wall of the bony canal,
united by transverse branches, and communicating
through the intervertebral foramina with the vertebral,
intercostal, lumbar, and sacral veins.

Strip off the pia mater from the chord; you find the
medullary substance divided into two lateral halves by
an *anterior and posterior fissure,* the posterior continued
from the calamus scriptorius: on each side of the middle
fissure is observed a *lateral groove,* on both surfaces of
the chord: these grooves are superficial, and from them
pass the anterior and posterior roots of the spinal nerves.

Within the vertebral canal in the neck, is seen the

Spinal Accessory Nerve, arising by small twigs from the side of the Spinal chord, as low down generally as the fourth or fifth cervical nerve: it then ascends between the ligamentum denticulatum and posterior roots of the three first cervical nerves, increasing in size by the addition of filaments, enters the foramen magnum, and passes forwards to accompany the par vagum.

OF THE SPINAL NERVES.

The Spinal Chord sends off twenty-nine pairs of spinal nerves. They consist of seven cervical, twelve dorsal, five lumbar, and five sacral pairs. Each of these nerves arises by two *fasciculi*, or bundles of filaments, one from the anterior, the other from the posterior surface of the chord : the ligamentum denticulatum intervenes between the two fasciculi, and they penetrate the dura mater by separate openings, receiving a tubular sheath from that membrane. The two fasciculi then pass outwards to the intervertebral foramen, the *posterior* swelling out into a small *oval* ganglion, within the foramen itself, and then immediately uniting with the *anterior fasciculus*, to form the nerve. Each nerve, thus formed, passes out of the spinal canal betwixt the vertebræ, and divides after a short course, into *two branches ;* one *anterior*, the other, *posterior*.

A general view of the distribution of the spinal nerves may be usefully subjoined in this place, to complete the description of the nerves.

1. The Cervical Nerves consist of seven pairs, emerging between the cervical vertebræ, the seventh nerve between the last cervical and first dorsal vertebra. The *posterior* branches pass backwards to the muscles and integuments behind the spinal column. The *anterior branches* of the first, second, and third cervical

nerves are interlaced with each other, and form the *Cervical plexus*, which is situated in the lateral part of the neck, and gives numerous ascending and descending branches. The large *anterior branches* of the four last cervical nerves and first dorsal unite to form the *Axillary* or *Brachial* plexus, which supplies the shoulder and upper extremity. The *Phrenic Nerve* is formed by filaments of the second, third, and fourth cervical nerves, and passes down the neck and through the thorax to the diaphragm.

. 2. The DORSAL NERVES are twelve pairs. Each nerve emerges betwixt the heads of the ribs, and divides into two branches, of which the *posterior* passes backwards between the transverse processes of the vertebræ, and is distributed to the muscles and integuments near the spine. The *anterior* branch is connected by a short twig to the nearest ganglion of the great sympathetic, runs outwards, covered by the pleura, to the angles of the ribs, and, passing between the two layers of the intercostal muscles, enters the groove in the lower margin of the rib, along which it accompanies the intercostal artery towards the anterior part of the chest. These anterior branches of the dorsal nerves are distributed to the muscles and integuments of the chest, giving also twigs to the muscles of the abdomen, and to the diaphragm.

· 3. The LUMBAR NERVES are five pairs, arising in the same manner, to escape by the intervertebral foramina of the lumbar vertebræ, and between the last vertebra of the loins and the upper surface of the sacrum. These nerves communicate with the great sympathetic nerve, and send small *posterior* branches to the muscles of the spine ; but the large *anterior branches* of the lumbar nerves assist in forming the *Crural Plexus*, from which proceed the great nerves of the thigh and leg.

4. The SACRAL NERVES are five or sometimes six in number on each side, arising from the inferior part or termination of the spinal chord. These nerves, with the two last lumbar pairs, descending within the spinal canal, produce the appearance, termed *Cauda Equina*. The sacral nerves differ from the other spinal nerves, in the situation of their ganglia and union of the two roots, by which each nerve arises, within the bony canal itself. The trunks thus formed divide into anterior and posterior branches, which pass out through the sacral foramina. The small *posterior branches* supply the neighbouring muscles; the large *anterior branches* come out through the anterior foramina, and unite with the lumbar nerves to form the crural plexus. The fifth sacral nerve escapes between the sacrum and coccyx, and the sixth, when it exists, through the notch of the latter bone, both contributing very little to the great plexus.

The *Crural Plexus*, which is frequently described as the *Lumbar* and *Sacral plexuses*, extends from the sides of the lumbar vertebræ into the pelvis: it gives off numerous lesser filaments to the parts within the pelvis, and in the perineum, and to the genital organs, and to the muscles and integuments of the abdomen, hip, and thigh: but it chiefly terminates in forming the great nerves of the lower extremity, viz. the *Anterior Crural Nerve*, the *Obturator Nerve*, and the *great Sciatic Nerve*.

All these nerves of the spine communicate freely by numerous twigs, and by the intervention of the GREAT SYMPATHETIC NERVE or INTERCOSTAL, which is now generally considered as belonging to a separate division of the nervous system, that of the GANGLIA. The great sympathetic nerve extends from the base of the skull, along the sides of the vertebræ, into the pelvis : it is first observed in the neck, where it forms the *Superior cervical ganglion*, close to the base of the skull, sending filaments into the carotid canal to communicate with the fifth and sixth cerebral nerves. It then descends through the neck, forming the *inferior cervical*, and sometimes a *middle cervical ganglion*, from which filaments pass to connect it with the other nerves, and to form plexuses about the heart and great vessels. In the thorax it runs down by the sides of the vertebræ, forming a ganglion between every two transverse processes, and communicating with the dorsal nerves; it here sends off the *greater* and *less splanchnic* nerves, which pass into the abdomen, the greater nerve terminating in the *Semilunar ganglion*, and *solar plexus*. In the abdomen, the great sympathetic is found descending on the sides of the lumbar vertebræ, and in the pelvis on the lateral anterior part of the sacrum, communicating with the lumbar and sacral nerves, and terminating on the os coccygis by the *ganglion impar*.

The Eighth Pair or Nervus Vagus has also a very long course; it arises in the head, and passes through the neck, to which it gives several branches. It enters the thorax anterior to the subclavian artery; here it gives off a remarkable branch, called the *Recurrent*, which is reflected back into the neck. The nerve then passes through the thorax, and entering the abdomen, terminates on the stomach; in this course, it has frequent communications with the great sympathetic, which it assists in forming the different plexuses that supply the thoracic and abdominal viscera.

CHAPTER VII.

DISSECTION OF THE ANTERIOR PART OF THE NECK.

SECTION I.

OF THE MUSCLES.

THIS is an intricate and important dissection, and re-
quiring the utmost care in its performance, when you
consider the variety of parts contained in the fore part
of the neck. The tubes which convey air to the lungs
and food to the stomach, the great vessels passing from
the heart to the brain and head, and the nerves emerging
at the base of the skull, are situated in the neck ; and
all these parts are embedded in cellular tissue, which
connects them, together with the muscles, closely.

The muscles of the anterior part of the neck, are sixteen in number on each side. They may be divided into
muscles situated superficially, muscles at the lower part
of the neck, and those situated at the upper part.

The SUPERFICIAL MUSCLES are two.

A block should be placed beneath the shoulders, and
the head thrown backwards, to make the region tense :—
then carry an oblique incision, through the skin, from
the chin to the middle of the clavicle, and cross its two
extremities by one incision along the line of the lower
jaw, and another in the direction of the clavicle ; raise
the integuments cautiously from the subjacent muscular
fibres, which are often pale and indistinct : immediately
under the integuments, and adhering to them, you
observe,

1. The MUSCULUS CUTANEUS, or *Platysma myoides.*

`—It *arises*, by slender separate fleshy fibres, from the cellular substance, covering the upper part of the deltoid and pectoral muscles. These fibres form a thin broad muscle, which runs obliquely upwards, and is

Inserted into the skin and muscles covering the lower jaw and cheek.

Use : To draw the skin of the cheek downwards, and, when the mouth is shut, to draw the integuments of the neck upwards.

Situation : This muscle should be dissected in the course of its fibres; it forms a thin layer, which extends from the upper part of the chest to the face, and is quite superficial. On dissecting off the skin, the fore part of the neck exhibits the appearance of being completely invested by a thin *fascia,* which is intermixed with the fibres of the platysma. This is named the *Cervical Fascia ;* above, it is seen descending from the angle of the jaw, cartilage of the ear, and surface of the Parotid gland ; below, it is continued with the cellular expansion covering the pectoralis major; its adhesions are very strong about the angle of the jaw, where it passes inwards between the adjacent portions of the Parotid and Submaxillary glands, and is connected with the stylo-maxillary ligament :—it ties down and invests the subjacent parts, also sending processes inwards among the muscles ; and it will be found connected with the sheath of the great vessels, and with a *deeper portion* of *Fascia,* which ascends from behind the clavicle and sternum.

Remove the platysma myoides from its origin, and invert it over the face. Immediately beneath it, is seen the *External Jugular Vein,* which is formed of branches from the temple, side of the face, and throat. It crosses obliquely over the sterno-mastoideus, passes behind the outer edge of that muscle, and dives beneath the clavicle, to enter the subclavian vein. Many small branches of the Cervical nerves ramify beneath the platysma ; and a large nerve, *the Nervus Superficialis Colli,* is observed turning over the posterior edge of the mastoid muscle, passing forwards to be distributed about the lower jaw, and sending off a considerable branch, the *N. auricularis,* behind the ear to the occiput. .

2. The Sterno-Cleido-Mastoideus— *Arises,* by two distinct origins, which are separated below by cellular tissue ; the *anterior,* tendinous and fleshy, and somewhat round, from the top of the sternum, near its junction with the clavicle ; the *posterior* or outer, fleshy and flat, from the upper and anterior part of the clavicle. These two origins soon unite, and form a strong muscle, which ascends obliquely upwards and outwards to be

Inserted, tendinous, into the outside of the mastoid process, and into the transverse ridge behind that process.

Situation : It is covered, in nearly its whole extent, by the platysma myoides ; its upper part is under the skin, except where it is lapped over by the parotid : it covers the parts on the side of the neck, and base of the skull.

Use : To turn the head to one side and rotate it ; when both muscles act, they bend the head forwards.

The muscle should be detached from the sternum and clavicle, and left suspended by its insertion ; or it may be divided across its middle. It is pierced by several branches of the cervical nerves, and at its upper part, nearly as high as the angle of the jaw, its inner surface is perforated by the *Nervus Accessorius.* These nerves ramify on the neighbouring muscles of the neck and shoulder. Between the posterior edge of the sternocleido-mastoideus and the fore part of the trapezius muscle, is seen a quantity of loose fatty substance, intermixed with the branches of the *Cervical* nerves. This fatty substance is watery and granulated, and is continued around the vessels under the clavicle. Numerous *absorbent glands* are met with in the side of the neck ; some of these are situated superficially, but the greater number are deep-seated, and are continued downwards behind the clavicle into the axilla.

The Anterior middle part of the throat is occupied by the *os hyoides*, placed at the base of the tongue, and by the *air-tube*, consisting of the *larynx* and *trachea*, passing downwards behind the sternum into the chest. These parts were seen, on removing the integuments, projecting between the receding bellies of the two sterno-mastoidei, and covered laterally by the fibres of the platysma myoides. The removal of the two superficial muscles exposes them more fully : they are covered by muscles, which pass to the jaw above, and to the sternum and shoulder below; and on each side of the trachea lie the great blood-vessels and nerves. You now perceive a *deeper-seated* portion of *Cervical Fascia*, which is attached to the clavicle and sternum, invests the hyoid and thyroid muscles, and passes inwards to the muscles of the spine. It closes the upper opening of the thorax, as will be seen on attempting to pass the finger downwards.

(1.) The Os Hyoides, or bone of the tongue, forms the uppermost of the projections beneath the chin ; it is divided into the *body* or *base*, and two *cornua* : the points at which each cornu is united to the body, are named the *Angles*, and here are situated two other smaller processes, the *lesser cornua*, or *appendices*.

(2.) The Larynx, or upper part of the trachea, situated immediately below the os hyoides, consists of

five cartilages, of which two are evident externally, viz,
1. The uppermost and largest is the *Thyroid Cartilage,*
which forms the projection termed *Pomum Adami;* it
is divided into two lateral parts or alæ, each of which
has a *superior* and *inferior cornu.* 2. The inferior is
the *Cricoid cartilage,* immediately below the thyroid,
and laterally in part covered by it. The two *arytenoid
cartilages,* and the *epiglottis,* lie behind this, and are
not at present seen.

(3.) The TRACHEA, consisting of cartilaginous rings,
extends from the cricoid cartilage into the thorax.

(4.) Behind the larynx is situated the PHARYNX. At
the part where the larynx terminates in the trachea, the
pharynx contracts itself, and forms the ŒSOPHAGUS,
which descends behind the trachea, situated rather to
the left side of the cervical vertebræ, and connected to
the vertebræ, and also to the trachea, by cellular tissue.

The MUSCLES at the lower part of the neck are five.

3. The STERNO-HYOIDEUS—*Arises,* thin and fleshy,
from the upper and inner part of the sternum, and from
the adjacent portions of the clavicle, and of the cartilage
of the first rib;—It forms a flat narrow muscle, which
ascends in the neck, approaching the muscle of the opposite side, and is

Inserted into the body or base of the os hyoides.

Situation : This pair of muscles is seen on removing
the platysma myoides, between the sterno-cleido mas-
toidei: there is visible in general, on the inner surface
of the sterno-hyoideus, a tendinous intersection, where
it is passing the thyroid gland.

Use : To draw the os hyoides downwards.

4. The OMO-HYOIDEUS—*Arises,* broad, thin, and
fleshy, from the superior costa of the scapula, near the
notch, and often from the ligament which converts the

notch into a foramen; it ascends across the neck, and forms a narrow middle tendon, where it passes below the sterno-cleido-mastoideus. Becoming fleshy again, it runs up, parallel to the last muscle, and is

Inserted into the base of the os hyoides, between its cornu and the insertion of the sterno-hyoideus.

Situation: This muscle is attached to the posterior border of the clavicle by the deep layer of cervical fascia, so that it does not proceed in a direct line from the scapula to the os hyoides :—its origin is covered by the trapezius; its middle part by the sterno-cleido-mastoideus; its anterior belly is seen on removing the platysma myoides. It crosses over the carotid artery and internal jugular vein.

Use: To draw the os hyoides obliquely downwards.

On dividing or holding aside the sterno-hyoideus, observe,

5. The STERNO-THYROIDEUS—This muscle *arises,* fleshy, from the inside of the sternum, and of the cartilage of the first rib; forms a flat layer, and is

Inserted into the inferior edge of the oblique ridge in the ala or side of the thyroid cartilage.

Situation: Beneath the sterno-hyoideus. It is a broader muscle; its posterior surface rests on the subclavian vein, lower part of the carotid, trachea and thyroid gland.

Use: To draw the thyroid cartilage, and consequently the larynx, downwards.

Under the sterno-thyroideus, we find the THYROID GLAND, a large reddish mass, situated on the superior rings of the trachea, and on the lateral parts of the larynx; in form somewhat like a crescent, with the cornua turned upwards. It is formed of two lateral lobes, or portions, connected by an intermediate transverse slip, or *isthmus*, which crosses the upper part of the trachea, just below the cricoid cartilage. This gland has no proper capsule, but is enveloped by dense cellular tissue; it is partly covered by the sterno-hyoid and thyroid muscles; and the isthmus or one of the lobes is sometimes prolonged as far as the os hyoides, by an ascending tongue-like process of the substance, or by a small slip of muscular fibres.

6. The THYRO-HYOIDEUS—*Arises*, fleshy, from the upper surface of the oblique ridge in the ala of the thyroid cartilage, and passes upwards, to be

Inserted into part of the base, and almost all the cornu of the os hyoides.

Situation: This short, thin, quadrilateral muscle appears like the continuation of the sterno-thyroideus: it is concealed by the sterno-hyoideus and omo-hyoideus, and is applied posteriorly to the thyroid cartilage and thyro-hyoid membrane.

Use: To draw the os hyoides downwards, or the thyroid cartilage upwards.

7. The CRICO-THYROIDEUS—*Arises*, tendinous and fleshy, from the side and fore part of the cricoid cartilage, and runs obliquely upwards.

Inserted, by two fleshy portions, the first into the lower part of the thyroid cartilage, and the second into its inferior cornu.

Situation: This is a thin muscle, placed on each side of the larynx, and covered by the sterno-thyroideus and thyroid gland.

Use: To pull forwards and depress the thyroid, or to elevate the cricoid cartilage.

The MUSCLES at the upper part of the neck comprise nine pairs,

8. The DIGASTRICUS—consists of two fleshy bellies,

and an intermediate tendon. The posterior belly *arises*, principally fleshy, from the fossa at the root of the mastoid process of the temporal bone; it descends obliquely forwards, and terminates in a strong round tendon, nearly two inches in length, which either perforates the fleshy belly of the stylo-hyoideus, or passes behind it. The tendon is then surrounded by a ring, or loop of tendinous fibres, which attach it to the upper border of the os hyoides, and which are of some length. It next changes its direction, and ascends, sending off from its inferior border a thin fascia, which passes over the surface of the mylo-hyoideus, and is also fixed to the base of the os hyoides. The muscle then becomes again fleshy, and runs obliquely upwards, approaching the muscle of the opposite side, to be

Inserted into a rough sinuosity on the inside of the lower jaw, close to the symphysis.

Situation : Its posterior belly is covered by the sterno-cleido-mastoideus ; its anterior belly lies immediately under the skin and platysma myoides ; the tendon has a small synovial capsule, where it is attached to the os hyoides.

Use : To depress the lower jaw ; or, when the mouth is closed, to raise the os hyoides, and with it the tongue and larynx.

In the triangular space formed by the two bellies of this muscle, and the base of the lower jaw, lies the SUBMAXILLARY GLAND : it lies upon a flat muscle, the mylo-hyoideus, which is seen between the two bellies of the digastricus. This gland is of an oval shape, and is invested by cellular tissue ; it is surrounded by several absorbent glands, and is covered anteriorly by the platysma myoides : posteriorly, it is sometimes continued with the parotid gland.

9. The STYLO-HYOIDEUS—*Arises*, tendinous, from the middle and inferior part of the styloid process of the temporal bone; its fleshy belly descends obliquely, and is generally perforated by the digastric tendon:

Inserted, tendinous, into the os hyoides at the juncture of its base and cornu.

Situation: This muscle is seen close to the posterior belly of the digastricus, being at its origin covered by that muscle, and situated more inwardly, and nearer the base of the cranium; it is the most superficial of three muscles which arise from the styloid process; sometimes it is accompanied by another small muscle, having the same origin and insertion, the *stylo-hyoideus alter*.

Use: To draw the os hyoides to one side, and upwards.

The two other muscles, which arise from the styloid process, may be traced at present: they are situated more inwardly.

10. The STYLO-GLOSSUS—*Arises*, tendinous and fleshy, from the lower half of the styloid process, and from the *stylo-maxillary ligament*, which connects that process to the angle of the lower jaw. It descends forwards and inwards, and becoming broader, but less thick, is

Inserted into the root of the tongue, runs along its side, and is insensibly lost near its tip.

Situation: This muscle lies within and rather above the stylo-hyoideus; underneath it, is a ligament, extending from the styloid process to the angle of the os hyoides:—This is the *Stylo-hyoid ligament*, and it frequently contains some ossified portions. The stylo-glossus spreads out at its insertion, and is intermixed with the hyo-glossus.

Use: To move the tongue laterally and backwards.

11. The Stylo-Pharyngeus—*Arises*, fleshy, from the root and inner part of the styloid process : it descends backwards and inwards, to be

Inserted into the side of the pharynx, and, by a few fibres, into the back part of the thyroid cartilage.

Situation: It is situated deeper and behind the styloglossus, and its fibres are intermixed with the constrictor muscles of the pharynx.—Close to the inferior border of this muscle we observe the *glosso-pharyngeal nerve.*

Use : To raise the pharynx and thyroid cartilage upwards.

On removing the submaxillary gland, and detaching the digastric muscle from the os hyoides and chin, we expose the next muscle.

12. The Mylo-Hyoideus—*Arises*, fleshy, from an oblique line on the inside of the base of the lower jaw, extending from the last dens molaris to the middle of the chin ; the fibres form a flat triangular muscle, converge, and are

Inserted into the lower edge of the base of the os hyoides.

Situation : This muscle unites with its fellow in a middle tendinous line, which extends from the os hyoides to the chin ; its posterior part is lined by the internal membrane of the mouth ; it lies under the digastricus, but is seen betwixt its bellies. Its lower and anterior part is covered by the thin fascia sent off by the digastricus, and is adherent to it.

Use : To draw the os hyoides forwards and upwards.

The *Lingual* or *Nerve of the ninth pair,* is observed coming from under the angle of the jaw, and passing behind this muscle. The Submaxillary gland also sends off a little process of its substance, together with its *ex-*

cretory duct, behind the posterior edge of the mylo-hyoideus : this duct * proceeds from the deepest part of the gland, it runs along the inner surface of the muscle forwards and upwards, on the inside of the sublingual gland, to open into the mouth, on the side of the fræ-num of the tongue, by a small projecting orifice.

The Sublingual Gland lies immediately above the mylo-hyoideus, betwixt it and the internal membrane of the mouth, where it lines the side and inferior surface of the tongue. It is the smallest of the salivary glands, and is placed beneath the anterior part of the tongue, where it can be readily felt by the finger ; it is sepa-rated from its fellow by the genio-glossi muscles. The lobules of this gland pour their secretion into the mouth by seven or eight minute orifices, between the root of the tongue and the side of the lower jaw.

The removal of the mylo-hyoideus exposes a pair of muscles, which are closely attached to one another, in their whole course from the chin to the os hyoides.

13. The *Genio-Hyoideus*—*Arises,* tendinous from a projection on the inside of the lower jaw, close to the symphysis ;—it descends, becoming broader, and is

Inserted into the base of the os hyoides.

Use : To draw the os hyoides forwards and upwards.

By removing the genio-hyoideus, or turning it back from its origin, we discover the next muscle.

14. The *Genio-Hyo-Glossus,* or *Genio-glossus*—*Arises*, by a short, but strong tendon, from a rough pro-tuberance on the inside of the lower jaw, higher up than the origin of the genio-hyoideus ; its fibres diverge in a very wide and radiated manner :—The superior and middle fibres pass horizontally to the under and lateral

---

* *Ductus Whartonianus.*

part of the tongue : the inferior, which are longer, descend obliquely backwards to the os hyoides.

Inserted into the whole inferior surface of the tongue, from the base to the apex; and into the posterior part of the body of the os hyoides, near its cornu.

Situation: This is a flattened triangular muscle, placed between the tongue and lower jaw : it lies under the genio-hyoideus before, and more outwardly under the mylo-hyoideus.

Use: According to the direction of its fibres, to draw the tip of the tongue backward into the mouth, the middle downwards, and to render its dorsum concave ; to draw its root and the os hyoides forwards, and to thrust the tongue out of the mouth.

15. The Hyo-Glossus is a broad flat muscle, situated more outwardly than the genio-glossus, and extending from the body and cornu of the os hyoides upwards to the inferior lateral part of the tongue. It *arises*, broad and fleshy, from the base, cornu, and appendix of the os hyoides ; the fibres pass directly upwards, to be

Inserted into the side of the tongue, near its root, where it mixes with the stylo-glossus.

Use: To move the tongue inwards and downwards.

16. The Lingualis consists of an irregular fasciculus of longitudinal fibres, placed beneath the root of the tongue laterally, and running forwards between the hyoglossus and genio-hyo-glossus, to be

Inserted into the tip of the tongue, along with part of the stylo-glossus.

Use: To contract the substance of the tongue, and bring it backwards.

OF THE VESSELS AND NERVES SEEN IN THE DISSECTION OF THE NECK.

I. ARTERIES.

THE CAROTID ARTERY * ascends from the thorax by the side of the trachea, to the hollow behind the angle of the jaw. It lies close upon the bodies of the vertebræ; on its outer side it has the *internal jugular vein*, with the *par vagum* lying between them, and behind is the *great sympathetic nerve*. The two great vessels with the par vagum, are enclosed and tied together by a dense cellular sheath; the intercostal or great sympathetic nerve lies close upon the spine, and is exterior to the sheath of the vessels. In the lower part of the neck, the carotid is covered by the sterno-cleido-mastoideus, and by the muscles running to the os hyoides and thyroid cartilage from the sternum and scapula; it here lies deep upon the muscles of the spine, and on the inferior thyroid artery. At the upper part of the neck, the artery is more superficial, being covered only by adipose tissue, absorbent glands, and by the platysma myoides. The *Descendens Noni nerve* is generally found on the fore part of the sheath of the vessels, sometimes within the sheath; and it forms, about the middle of the neck, a delicate *plexus* with some branches of the cervical nerves:—behind the artery, near the trachea, we also find the *Nervus superficialis cordis*, a slender filament from the superior cervical ganglion: and many absorbent glands are found along the artery in its whole course.

The Carotid gives off no branches, until it reaches the space between the larynx and angle of the jaw: here, just below the cornu of the os hyoides, it divides into the *External* and *Internal Carotids.*— Of the two, the INTERNAL CAROTID is at first situated more outwardly; it is then seen ascending behind the external carotid, in

* *Primitive or Common Carotid.*

front of the cervical transverse processes; it passes deep to the base of the cranium, not giving off any branch, and entering the foramen caroticum, turns along its winding canal into the skull.

The EXTERNAL CAROTID is placed anterior to the internal, and nearer to the larynx; from the side of the larynx, it ascends behind the angle of the lower jaw, inclining backwards, and passing under the posterior belly of the digastricus and stylo-hyoideus, and crossed by the nerve of the ninth pair. It then holds its course directly upwards between the ascending plate of the lower jaw and the ear, being concealed beneath the parotid gland; and it here terminates, at the neck of the jaw, by bifurcating into the internal maxillary and temporal arteries.

BRANCHES OF THE EXTERNAL CAROTID.

These are eight in number, subject to some variety in the mode of coming off: they may be distinguished into *anterior* and *posterior* branches.

Anteriorly, the external carotid sends off

1. The SUPERIOR THYROID ARTERY:—this artery usually comes off from the carotid close to its origin. It descends obliquely, in a winding sinuous course, under the omo-hyoideus and sterno-thyroideus muscles, to the upper part of the thyroid gland. Its *branches* are (1.) Small *muscular* arteries. (2.) The *laryngeal artery* which passes transversely behind the thyro-hyoideus muscle, accompanying the laryngeal nerve, and penetrates between the os hyoides and thyroid cartilage, to the muscles and internal parts of the larynx. (3.) *Ramus crico-thyroideus*, a smaller artery, frequently a branch of the laryngeal, which ramifies over the crico-thyroid membrane. (4.) The large terminating or *thy-*

roid branch ramifies on the thyroid gland, inosculating with the corresponding artery, and with the inferior thyroid.

The thyroid artery is variable in size: it sometimes gives a large branch to the sterno-mastoid muscle. The thyroid has been found arising from the Carotid trunk.

2. The LINGUAL ARTERY is the next branch; it comes off from the carotid behind the digastric muscle, and runs forwards and inwards toward the os hyoides. It sinks under the hyo-glossus muscle, lying upon the middle constrictor of the pharynx, and passing above the cornu of the os hyoides: it then bends upwards between the hyo-glossus and genio-glossus, and between this last muscle and the sublingual gland, to the base of the tongue. Here, taking the name of the *Ranine Artery*, it runs horizontally forwards, beneath the middle part of the tongue, to its apex, lying close to the fraenum. From the carotid to the os hyoides, the Lingual artery is accompanied by the Lingual nerve, which is then separated from the artery by the hyo-glossus muscle. Its branches are (1.) To the muscles under the base of the jaw. (2.) *Dorsalis linguae*, to the base of the tongue, epiglottis and palate. (3.) The *Sublingual branch* to the sublingual gland, and parts adjacent. (4.) Numerous lateral branches to the substance of the tongue from the *A. ranina*, which passes forwards, accompanied by the terminating branch of the lingual nerve, and anastomoses with its fellow by an arch at the tip of that organ.

3. The EXTERNAL MAXILLARY or FACIAL ARTERY is a large branch, arising either by a common trunk with the lingual, or immediately after it. It passes forwards under the lower jaw, behind the stylo-hyoideus and digastric muscles, and then buries itself under the sub-

maxillary gland. It is here very tortuous, and is next seen advancing forwards, becoming superficial, and bending over the lower jaw, immediately before the anterior edge of the masseter muscle : it now becomes the proper artery of the face, and its further course will be described in the dissection of that part. The branches of the external maxillary are (1.) The *inferior palatine branch* to the lateral part of the pharynx, tonsils, and palate. (2.) Several *branches* to the submaxillary gland and muscles above the os hyoides. (3.) The *Submental artery*, which runs along the line of the lower jaw, between the mylo-hyoideus and anterior portion of the digastricus, and, turning over the jaw near the symphysis, ramifies on the muscles and integuments of the under lip.

Posteriorly, the external carotid sends off

4. The OCCIPITAL ARTERY :—this artery is of considerable size, it comes off nearly opposite to the lingual, and passes to the occiput. It ascends obliquely backwards, crossing over the internal carotid artery and internal jugular vein, and running beneath the sterno-cleido-mastoideus, and along the posterior belly of the digastricus; it passes in the space between the transverse process of the atlas and the mastoid process of the temporal bone, is lodged in a superficial groove behind the mastoid process, and then curves backwards on the occipital bone, covered by the splenius muscle. Emerging from the posterior margin of the splenius, it runs upwards in a tortuous manner upon the posterior part of the head, inosculating with the other arteries of the scalp. Its branches are (1.) *Muscular* branches. (2.) A *Meningeal* artery, through the foramen lacerum posterius. (3.) Some *descending* tortuous branches, which pass to the base of the skull, and inosculate, near the foramen

magnum, with ramifications of the vertebral and posterior cervical arteries.

5. The ASCENDING PHARYNGEAL ARTERY is small and
deeply seated; it passes inwards to the pharynx and
base of the skull: it also sends into the cranium some
small *Meningeal* arteries. This artery is sometimes
double, or it is frequently given off by the occipital.

6. The ARTERIA POSTERIOR AURIS, or *Auricular artery*, is also a small artery, coming off from the carotid
under the parotid gland. It passes upwards between the
ear and mastoid process, bifurcating into two branches,
of which (1.) The *anterior* ramifies on the convex surface
of the cartilage of the ear, (2.) The *posterior* ascends
over the mastoid process to the side of the head. This
artery sends a *stylo-mastoid* branch to the internal ear.

The terminating arteries of the external carotid are
7. and 8. The TEMPORAL and INTERNAL MAXILLARY
ARTERIES, which will be described in the dissection of
the face.

II. VEINS.

The INTERNAL JUGULAR VEIN is a continuation of
the lateral sinus, which is prolonged into the vein through
the foramen lacerum in the base of the cranium. This
great vein is deeply seated in the whole of its course; it
begins at the base of the cranium, in the jugular fossa,
by an *ampulla* or dilatation, sometimes termed the
sinus of the jugular vein; it is here placed at first close
to the internal carotid, and then comes out from under
the angle of the jaw, and from behind the styloid process and styloid muscles, to take its course down the
neck on the outer side of the carotid trunk; it lies,
together with the par vagum, in the same sheath with
the artery, which vessel it partly overlaps :—before it

6

reaches the thorax, it passes rather more forward than the artery to join the subclavian vein.

The branches of the Internal Jugular correspond generally with those of the external carotid; above the larynx, it receives the facial, lingual, pharyngeal, and occipital veins, and has also a considerable branch of communication with the external jugular; lower down, it is joined by the superior thyroid vein, and then, in descending, usually receives some middle thyroid veins.

The EXTERNAL JUGULAR VEIN is the sub-cutaneous vein of the neck. It arises by branches in the Parotid region, chiefly from the internal maxillary and temporal veins, appearing as a trunk at the angle of the jaw, under the parotid, and here receiving the posterior auricular vein. It was seen in the dissection of the muscles, crossing the sterno-mastoideus, and then sinking behind the outer border of the muscle, to enter the subclavian vein, more outwardly than the internal jugular: it receives cutaneous cervical veins, and transverse scapular branches from the shoulder, some of which usually form a considerable plexus above the clavicle.

III. NERVES.

1. The NERVE of the EIGHTH PAIR, PAR VAGUM or *Nervus Vagus*, is discovered on separating the internal jugular from the trunk of the carotid, lying in the same sheath with these vessels, but rather behind them : its precise situation is on the outer side of the artery, and behind the great vein, to which it is more closely adherent. On tracing it upwards, it is found deeply seated at the base of the skull, coming out of the foramen lacerum with the jugular vein ; it is continued down the neck between the two great vessels, and then passes into the thorax. In this course it gives off several branches.

(1.) At the base of the cranium, the Nervus Vagus

swells out into a sort of *ganglion,* and sends off *filaments* which are connected with the other nerves at the base of the skull, as the lingual nerve, glosso-pharyngeal, spinal accessory, and the upper cervical ganglion.

(2.) The *Pharyngeal Branch,* of some size, passes inwards to the pharynx, where, with filaments of other nerves, it forms the *pharyngeal plexus,* on the constrictor muscles.

(3.) The *Superior Laryngeal Nerve* is a larger branch; it descends obliquely behind the internal carotid artery, and subdivides into *(a,)* The *External Laryngeal Nerve,* which ramifies on the outside of the larynx, and *(b,)* The *Internal Laryngeal,* which pierces the hyo-thyroid membrane, accompanying the *laryngeal artery,* to supply the internal parts of the larynx.

(4.) Some *Cardiac* filaments pass off in the lower part of the neck, and descend along the carotid to join the cardiac plexuses.

The Nervus Vagus having reached the lower part of the neck, enters the thorax behind the subclavian vein, passing on the right side before the subclavian artery, and, on the left, before the arch of the aorta.

2. The INTERCOSTAL or GREAT SYMPATHETIC NERVE, is found descending behind the carotid, in the cellular tissue between that vessel and the muscles covering the vertebræ. It is distinguished from the nervus vagus by being smaller, lying nearer to the trachea, and adhering to the muscles of the spine; also by its forming several ganglia.

The first ganglion is the *Superior Cervical;* this, which may be considered as the beginning of the nerve in the neck, is situated behind the internal carotid, under the base of the skull, in front of the rectus anticus muscle; it is of a reddish colour and spindle-like shape, varying in length, but generally extending from the orifice of the carotid canal to the third cervical vertebra. This ganglion gives off two superior filaments, which ascend through the carotid canal to join the fifth and sixth cerebral nerves. The Sympathetic nerve then contracts itself, and, about the fifth cervical vertebra, enlarges again into the *Middle Cervical Ganglion*, which is often small, and sometimes wanting. The nerve then descends close to the spine ; and, opposite to the seventh cervical vertebra and neck of the first rib, forms the *Inferior Cervical Ganglion*, which will be examined in the thorax.

The branches of the great sympathetic are numerous, generally passing off from the ganglia. At the base of the cranium, twigs communicate with the facial nerve, eighth and ninth pairs, and upper cervical nerves :— from the superior cervical ganglion the *Nervi molles*, so denominated from their soft texture, ramify over the carotid and its branches; and a single filament, the *Nervus Superficialis cordis*, descends behind the artery, near the trachea : filaments also pass to the larynx, pharynx, and muscles of the spine.—Other *Cardiac filaments* also pass from the middle and inferior cervical ganglia, and from the union of these cardiac nerves, with filaments of the par vagum, extensive plexuses are formed over the base of the heart, and on the great vessels.

3. The GLOSSO-PHARYNGEAL NERVE has been seen within the skull, as one of the divisions of the eighth pair, passing with the par vagum through the foramen lacerum. In the neck it lies deep under the angle of the jaw, and, at the base of the cranium, will be found separated from the par vagum by the trunk of the internal jugular vein ; it directs its course downward and forward between the internal and external carotid arte-

ries, and is seen close to the inferior border of the stylo-pharyngeus; it passes between that muscle and the stylo-glossus, to the under and back part of the tongue. It gives filaments to the stylo-pharyngeus and constrictor muscles of the pharynx, and is also connected with the other nerves, and then passing under the stylo-glossus and hyo-glossus muscles, is distributed largely to the tongue, chiefly supplying the mucous membrane.

4. The NINTH PAIR, HYPOGLOSSAL or LINGUAL NERVE, having emerged from the anterior condyloid foramen, is connected at the base of the skull with the eighth pair and sympathetic. Like them, it lies deep, and comes out from under the angle of the jaw. It is seen passing from behind the internal jugular vein, and crossing over both carotids, being covered by the stylo-hyoideus and digastric muscles. Having descended as low as the angle of the jaw, it curves upwards, beneath the tendon of the digastricus, sending off from its convexity the *Descendens Noni*, and then passes forwards to the side of the tongue. This lingual nerve is at first in contact with the lingual artery, but is afterwards separated from it by the hyo-glossus muscle; it passes under the mylo-hyoideus between that muscle and the hyo-glossus, spreading out, and giving filaments to the muscles of the pharynx, os hyoides and tongue. Its branches are (1.) *Nervus Descendens Noni*, which descends over the great vessels, and terminates in the long anterior hyoid and thyroid muscles. (2.) *Muscular* branches. (3.) The terminating branch or trunk of the lingual nerve, on reaching the anterior border of the hyo-glossus, plunges between the genio-glossus and lingualis muscles, joining the *Ranine artery*, and runs forwards under the tongue as far as within an inch of the apex, being distributed to the muscular fibres.

5. The LINGUAL or GUSTATORY BRANCH, sent off by the INFERIOR MAXILLARY or *third branch* of the *fifth pair of nerves,* is also seen in the neck, descending from the inside of the lower jaw. It is found under the mylo-hyoideus, close upon the lower edge of the jaw-bone, emerging from the inferior border of the pterygoideus internus. The nerve then passes, with the submaxillary duct, between the mylo-hyoideus and hyo-glossus, and turns upwards above the sublingual gland, to the lateral part of the tongue. It gives twigs to the submaxillary and sublingual glands, and parts adjacent, and terminates by numerous filaments, which are finally distributed to the mucous membrane of the tongue.

6. The SPINAL ACCESSORY NERVE *, having passed out of the cranium with the par vagum, separates from it, and descends behind the internal jugular vein; it turns outwards to reach the sterno-cleido-mastoideus, perforates that muscle, (or sometimes passes beneath it,) and then emerging from its posterior margin, terminates in the trapezius, being much connected with the cervical nerves.

7. The SEVEN CERVICAL NERVES come out from the foramina betwixt the vertebræ. The *posterior branches* of these nerves pass backwards to the muscles of the spine; the *anterior branches,* passing downwards along the vertebræ, unite intricately with each other. The CERVICAL PLEXUS is formed by the anterior branches of the three upper cervical nerves; it is found by the sides of the corresponding vertebræ, resting on the scaleni and levator scapulæ muscles, beneath the posterior border of the sterno-mastoideus. This plexus is intermixed with vessels, lymphatic glands, and fat: it sends off many branches,—downwards to the shoulder, chest, and axilla;

* *Superior External Respiratory Nerve* of Mr. Bell.

upwards, to the lateral part of the head ; and, anteriorly, branches over the surface of the mastoid muscle.

The *Anterior* branches of the four lower Cervical Nerves, and of the first Dorsal, form the GREAT BRACHIAL or AXILLARY PLEXUS, which is seen in the lower part of the neck, accompanying the subclavian artery, beneath the clavicle, into the axilla.

The PHRENIC NERVE * is also seen in this dissection, coming from the cervical plexus, and chiefly formed by filaments of the third, fourth, and fifth cervical nerves. This small nerve lies upon the fore part of the anterior scalenus muscle, and dives into the thorax between the subclavian artery and vein.

8. The RECURRENT NERVE, or *Inferior Laryngeal,* a branch sent off from the par vagum in the thorax, is also seen in the neck.   Dissect between the under surface of the trachea and œsophagus in the lower part of the neck, and you will find the recurrent nerve situated there ; it is distributed to the œsophagus, trachea and larynx.

CHAPTER VIII.

DISSECTION OF THE THORAX.

SECTION I.

OF THE MUSCLES WHICH LIE UPON THE OUTSIDE OF THE THORAX.

IN taking off the integuments from the chest, the broad pectoral muscle covering it, and neighbouring edge of the deltoid must be dissected in the course of their fibres, which run obliquely from the sternum and clavicle to the

* *Internal Respiratory Nerve* of Mr. Bell.

upper part of the humerus. Carry the arm forcibly from the body, to make these fibres tense, and direct an incision from the sternal portion of the clavicle to the front edge of the axilla, and dissect off the integuments in this line; or the skin may be reflected from below upwards in the line of the lower border of the muscle, from the cartilage of the fifth rib to the shoulder.

These muscular fibres are not covered by any fascia, but by a condensed cellular membrane, which is similar to a thin aponeurosis, and is with difficulty dissected away, unless it be done very regularly. Beneath the integuments in the female, we find the Mammary Gland, which is of a soft texture, of a flattened oval figure, lobulated, and embedded in adipose substance: the gland is invested by a dense *cellular tunic*, which binds the lobules together, and sends processes between them. It is loosely united by cellular tissue to the surface of the pectoralis major; in the centre of the skin covering it, is the Nipple or *Papilla*, with the *areola* of dark colour encircling it: at the apex of the nipple, the numerous orifices of the *lactiferous ducts* may be observed.

Three pairs of muscles are described in this dissection.

1. The Pectoralis Major—*Arises*, by short tendinous fibres, from the inner or sternal half of the clavicle; tendinous, from the anterior surface of the sternum, its whole length; and, by fleshy fibres, from the cartilages of the seven true ribs, except the first. The muscle thus formed consists of two portions, the *thoracic* and the *clavicular*, separated by a line of cellular tissue. The fleshy fibres run obliquely across the breast, and, converging, form a strong flat tendon, which is

Inserted into the anterior margin of the bicipital groove in the humerus.

Situation: The belly of the muscle is superficial. It

is separated from the deltoid by cellular tissue, the *cephalic* vein, and by the *Humeral-thoracic* artery. Its tendinous fibres, arising from the sternum, are interlaced with those of the opposite muscle, so as to form a kind of fascia covering the bone; and its lower border inter-mixes with the external oblique, sometimes by a distinct fleshy slip. The tendon is covered by the front edge of the deltoid; it forms the anterior fold of the arm-pit, and appears twisted, for the fibres, which proceed from the thoracic portion of the muscle, seem to pass behind those proceeding from the clavicle, and to be inserted into the os humeri somewhat higher up.

Use : To move the arm forwards and upwards, towards the sternum; or to depress it, when elevated : when the arm is fixed, to raise the ribs and trunk of the body.

The pectoralis major should be lifted from its origin, and suspended by its tendon; vessels and nerves from the axilla are seen piercing its under surface.—This will expose the next two muscles.

2. The Pectoralis Minor—*Arises,* by three tendinous and fleshy digitations, from the upper edges of the third, fourth, and fifth ribs, near their cartilages; it forms a fleshy triangular belly, which becomes thicker and narrower as it ascends, and is

Inserted, by a short flat tendon, into the anterior part of the coracoid process of the scapula.

Situation : This is a much smaller muscle; its belly is covered by the pectoralis major; the tendon * passes under the anterior edge of the deltoid, and is connected at its insertion with the origins of the coraco-brachialis, and of the short head of the biceps flexor, and also with

---

* In some subjects, the tendon slides over the coracoid process, and spreads out widely, adhering to the root of that process, to the tendon of the supra-spinatus, and capsule of the shoulder.

the strong ligament, which passes from the external edge of the coracoid process to the acromion. The upper digitations of the serratus magnus are observed arising from the ribs behind the pectoralis minor.

Use : To draw the scapula forwards and downwards, and, when that bone is fixed, to elevate the ribs.

The next muscle is situated under the clavicle, and is not at first perceived ; it is concealed by the bone, and is covered by a thin fascia.

3. The SUBCLAVIUS—*Arises*, by a flat tendon, from the cartilage of the first rib, and forms a broad fleshy belly, which is

Inserted into the inferior surface of the clavicle, from near its sternal extremity as far as its ligamentous connexion with the coracoid process.

Situation : This muscle is concealed by the pectoralis major, and anterior part of the deltoid : the thin *fascia* * covering it is of considerable extent, descending from the clavicle and coracoid process to the upper ribs, and arching over the axillary vessels.

Use : To draw the clavicle and shoulder downwards.

Having lifted up the pectoralis minor from its origin, the situation of the subclavian vessels, which pass under the clavicle, and over the first rib, may be seen. (See the dissection of the axilla.)

SECTION II.

OF THE PARTS CONTAINED WITHIN THE CAVITY OF THE THORAX.

THE cavity of the thorax may now be opened, by cutting through the cartilages of the ribs on each side, and

---

* *Costo-coracoid membrane.*

separating the lower part of the sternum from the diaphragm. That bone must then be lifted upwards, and
removed, by separating it at its articulations with the
clavicles.

On looking under the sternum, while it is lifted up,
we see the *Mediastinum*, separating, as it is gradually
torn from the posterior surface of the sternum, into
two layers, and thus forming a triangular cavity. This
cavity is in a great degree artificial, and is owing to the
method of raising the sternum.

When the sternum is laid back or removed, the following parts are to be observed :

The MEDIASTINUM, now collapsed, dividing the thorax
into two distinct cavities, of which the right is the largest.

The LUNGS of each side lying distinct in these cavities.

The PERICARDIUM, containing the heart, situated in
the middle of the thorax, between the two laminæ of the
mediastinum, and protruding into the left side.

The internal surface of the PLEURA, smooth, colourless, and glistening, lining the ribs, and reflected over
the lungs.

1. The PLEURA belongs to the class of serous membranes, forming a shut sac, without an aperture. Each
side of the thorax has its particular pleura.—The pleuræ
are like two bladders, situated laterally with respect to
each other; by adhering together in the middle of the
thorax, and passing obliquely from the posterior surface
of the sternum to the dorsal vertebræ, they form the
mediastinum.—The pleura lines the ribs, and the upper
surface of the diaphragm, and is reflected over the lung,
which is in fact behind it; it forms the *ligamentum latum pulmonis*, a reflection of this membrane, which connects the inferior edge of the lung to the spine and
diaphragm.

The Pleura, where it lines the ribs and intercostal muscles, is termed *Pleura Costalis*, in contradistinction from the *Pleura Pulmonalis* investing the lungs. The attachment of the two pleuræ to the sternum is generally oblique, so that, in passing downwards, the mediastinum inclines towards the left side, especially near the diaphragm. In description, the mediastinum is divided into the anterior and the posterior. The *Anterior Mediastinum* is that portion extending from the front of the lungs and sides of the pericardium to the sternum ; it contains some lymphatic glands, and adipose tissue, the internal mammary, vessels, and, at its upper part, the thymus gland. The *Posterior Mediastinum* extends from the roots of the lungs to the spine, and contains important parts to be described hereafter. In the *middle space*, between the anterior and posterior mediastinum, are placed the *heart* and *pericardium*.

2. The Lungs—*Colour*, reddish in children, greyish in adults, and bluish in old age.—*Shape*, corresponding to that of the thorax, somewhat pyramidal. Each lung has an *external surface*, smooth and convex towards the ribs ; an *internal surface*, flattened and corresponding to the mediastinum ; a *posterior* rounded border, lodged in the hollow by the side of the vertebral column ; an *anterior border*, sharp and thin : a *base*, concave and elongated, corresponding with the diaphragm ; and, finally, a conical *summit* or *apex*, which ascends, in a cul-de-sac of pleura, somewhat above the level of the first rib.

Division (1.) The Right Lung is the largest, and is divided into three lobes, two greater ones, and an intermediate lesser lobe.

(2.) The Left Lung has two lobes, and also a square notch or depression opposite the apex of the heart ;

into the deep fissures which form the divisions of the lungs into lobes, the pleura enters; that part of the lung which is affixed to the spine, is called its *root;* it is the part by which the bronchus, or division of the trachea, and the great pulmonary vessels enter.

3. The PERICARDIUM is a strong, white, *fibro-serous* membrane, smooth and lubricated upon the inside, forming a bag for containing the heart, and having its inner serous lamina reflected over the roots of the great vessels and substance of the heart itself. It is situated behind the sternum, between the two pleuræ, and, below, it adheres strongly to the central tendon of the diaphragm.

4. When you slit open the fore part of the pericardium, you expose the HEART, lying loose in the cavity, and unattached except at its base, where it is fixed by its great vessels and by the reflection of the pericardium.—Observe, that the heart is placed obliquely in the thorax, with its base directed upwards and towards the right side, while its apex points downwards and to the left side, so that, in the living body, it is felt striking between the fifth and sixth ribs. Its shape is conical; its anterior or upper surface convex, the posterior flattened, where it rests upon the diaphragm. The base or broad upper part is occupied by two cavities, named AURICLES; the lower, thicker, and more conical part consists of the two VENTRICLES : a *transverse depression* separates the auricles from the ventricles, while an *oblique groove* both on the anterior and posterior surfaces, occupied by the coronary vessels, marks the situation of the *septum cordis,* or division between the cavities of the right and left side.

Examine the situation of the great cavities of the heart, and of the vessels within the pericardium. The

VENA CAVA SUPERIOR is seen coming down from the upper angle of the pericardium. The INFERIOR CAVA is seen coming up through the diaphragm, but only a very small part of this vein is covered by the pericardium, while the cava superior has a course, within the bag, of nearly two inches : the two veins enter the right auricle. The RIGHT AURICLE is placed at the right side of the base of the heart, extending forwards with a pointed anterior *appendage*, which is seen between the aorta and right ventricle : this auricle might be called the anterior; it generally appears black, from the blood shining through its thin coats. The RIGHT VENTRICLE is also situated anteriorly, forming the right border of the heart ; it feels loose, less resisting, and partly wrapt round the left ventricle. The PULMONARY ARTERY arises from the right ventricle; it ascends on the left side of the aorta, and, after a course of about two inches, divides, while still within the pericardium, into (1.) The *right pulmonary Artery*, which passes under the arch of the aorta, crosses behind it and the vena cava superior to the right lung, and is the longest; and, (2.) the *left pulmonary Artery*, which passes to the left lung, crossing the descending aorta anteriorly. The PULMONARY VEINS enter the left auricle, two veins come from each lung : the right veins are longest, as they pass behind the vena cava superior. The LEFT AURICLE is situated on the left side of the right auricle, and somewhat behind it; its tip, or *auricular appendage*, is seen lapping round upon the LEFT VENTRICLE : this is situated behind and on the left side of the right ventricle ; its substance is stronger and more firm to the touch, and it forms the rounded left border and apex of the heart. The AORTA arises from the back part and right side of the left ventricle ; its root is covered by the pulmonary artery; but it soon emerges from beneath it, and then ascends betwixt that artery and the vena cava superior. Immediately from the root of the aorta, within the pericardium, the two *Coronary Arteries* are sent off to supply the heart itself.

SECTION III.

DISSECTION OF THE GREAT VESSELS OF THE HEART.

Now proceed to examine the situation of the *Great Vessels* of the *Heart* :—the sternum having been removed, remark, immediately behind its upper part, the large *Venous trunks* passing from the neck and upper extremities to the right side of the heart; more deeply seated, and rather below, the *Arch* of the *Aorta*, with its great branches, and, behind the aorta, the *Trachea* descending from the neck into the thorax.

First observe the VENA CAVA SUPERIOR; it will be seen descending before the root of the lungs, and on the right side of the aorta. Immediately before perforating the pericardium, it is joined upon its posterior part by the vena azygos, which comes forwards from the spine, returning the blood from the intercostal spaces.

Behind the right margin of the Sternum, and just above the arch of the aorta, the superior cava is seen dividing into two great branches *.

1. A *short branch,* coming obliquely downwards from the *right* side, formed by the RIGHT SUBCLAVIAN VEIN, and the RIGHT INTERNAL JUGULAR.

2. A *longer* and *larger trunk,* from the *left* side, formed by the LEFT SUBCLAVIAN, and LEFT INTERNAL JUGULAR: this venous trunk crosses in a horizontal

---

* *Venæ Innominatæ* of many writers.

7

direction, before the trachea and arteries arising from the arch of the aorta; it is situated immediately above the arch of the aorta, and behind the sternum, from which it is only separated by loose cellular tissue.

On each side, the internal jugular vein descends along the neck by the side of the carotid, while the subclavian vein comes from the arm. Into the posterior part of the angle formed by the union of the left subclavian and left jugular, the *Thoracic Duct* empties itself.

The VENA CAVA INFERIOR, immediately after passing through the diaphragm from the abdomen, enters the pericardium.

The AORTA leaves the heart opposite the fourth dorsal vertebra; it crosses over the pulmonary artery, ascends obliquely upwards, and to the right side, as high as the second dorsal vertebra. Here it forms an ARCH or incurvation, which passes from the right to the left side, and at the same time obliquely from before backwards; it then comes in contact with the upper part of the third dorsal vertebra, and descends along the left side of the spine in the posterior mediastinum. This arch of the aorta is situated behind the first bone of the sternum, behind and somewhat below the left branch of the cava superior, and immediately before the trachea, just above its bifurcation into the two *bronchi*.

From the upper part of the arch come off three large arteries.

1. The ARTERIA INNOMINATA, or common trunk of the right carotid and subclavian, ascends obliquely over the trachea, to the right side, and, after a course of about an inch, arriving at the level of the sterno-clavicular articulation, bifurcates into

(*a,*) The RIGHT CAROTID, which ascends in the neck by the side of the trachea.

(b,) The RIGHT SUBCLAVIAN, which passes outwards to the arm.

2. The LEFT CAROTID is the second large artery from the arch of the aorta; it comes off close to the root of the Innominata, and passes over the left edge of the trachea, close to the œsophagus, to take its place in the neck.

3. The LEFT SUBCLAVIAN comes off from the extremity of the arch, and is situated deep in the chest.

Some varieties occur in the arteries from the arch of the aorta. The right carotid and right subclavian may arise separately, sometimes the latter springs from the extremity of the arch, and passes between the trachea and œsophagus, to reach its place on the right:—or the arch may give off the vertebral or inferior thyroid arteries.

The THYMUS is a soft glandular body, lying before the lower part of the trachea and great vessels of the heart, in the uppermost part of the anterior mediastinum. It is very large in the fœtus, smaller in adults, and nearly disappears in the aged.

Where the aorta begins to descend, it is connected to the pulmonary artery by a ligament, which in the fœtus was a large canal, the DUCTUS ARTERIOSUS.

COURSE OF THE SUBCLAVIAN ARTERY.

The subclavian artery, on each side, passes upwards and outwards, from the chest, over the middle of the first rib; it crosses the rib between the bellies of the anterior and middle scaleni muscles, passes beneath the clavicle and subclavius muscle, and then, assuming the name of AXILLARY ARTERY, inclines downwards under the arch of the pectoralis minor, to enter the hollow of the axilla.

To examine this course of the subclavian artery, the sternal extremity of the clavicle may now be detached from the ribs, and turned outwards *. The great artery, before passing the scaleni, is deeply-seated, being immediately adjacent to the bag of the pleura ; it is covered, anteriorly, by the great venous trunks, and by the mastoid and sterno-hyoid and thyroid muscles : the nervus vagus passes before it into the chest, and behind is the great sympathetic nerve, encircling it with the filaments of the inferior cervical ganglion.—Where it is passing over the first rib, the artery, although still placed deeply, is separated from the integuments only by cellular tissue, by the deep cervical fascia, and fibres of the platysma ; it is closely applied to the rib, lying on its groove or flattened surface, immediately behind the *tubercle*, to which the anterior scalenus is fixed ; and it passes under the clavicle at the distance of rather more than one-third of the length of the bone from its sternal extremity.— The Nerves of the Axillary plexus are behind and above the artery, when it is emerging from the scaleni : the subclavian vein is situated in front of the artery, but is separated from it by the anterior scalenus.

But the diversity of origin of the artery on the right and left side, occasions some important differences in the manner, in which the two subclavians emerge from the chest.

The Right Subclavian is in general rather larger than the Left, and is also shorter ; in situation the right is more superficial ; this is owing to the oblique direction of the arch of the aorta. Both arteries form a curve but while the right subclavian passes obliquely outwards

---

* This will interfere with the dissection of the axilla, which may be examined, therefore, previously.

to the interval between the two scaleni, the left ascends vertically, and suddenly bends outwards to reach the space between these muscles. The right subclavian is separated from the spinal column and longus colli muscle by an interval of some extent, while the left is applied closely to these parts, and ascends on the left side of the œsophagus, behind the thoracic duct and left jugular vein. On the right side the Recurrent nerve curves round the subclavian, on the left side it turns round the arch of the aorta.

Having reached the scaleni, the great Artery on each side has exactly the same relations. It takes the name of Axillary, as soon as it has fairly passed beyond the lower border of the rib.

BRANCHES OF THE SUBCLAVIAN ARTERY.

These are subject to much variety, in their order of coming off,—in their number, sometimes arising singly, at other times in large trunks, which subdivide,—and also in their volume, since the chief branches are not always derived from the same trunks. In general, they are sent off from the subclavian close together, where it is crossing the root of the neck to reach the scaleni.

Upwards the subclavian sends off

1. The VERTEBRAL ARTERY.—This artery is usually the first and largest branch, arising from the back part of the subclavian; it lies deep in the neck, passes upwards behind the inferior thyroid artery, over the longus colli muscle, and enters the foramen in the transverse process of the sixth or seventh cervical vertebra. It ascends through the canal formed by the transverse processes, bends outwards to pass through the transverse process of the atlas, runs horizontally in a groove on the upper surface of that vertebra, and then enters the fora-

men magnum of the occipital bone.—In this course it gives twigs to the muscles and adjacent parts, and it is accompanied by a large vein.

2. The INFERIOR THYROID ARTERY comes off from the subclavian, further out and in front of the vertebral: it ascends at first directly upwards, within the anterior scalenus and upon the longus colli muscle; it then turns inwards, passing transversely behind the carotid artery, and reaches, in a tortuous course, the thyroid gland; it divides into branches, which penetrate the under part of the gland, and inosculate largely with the artery of the other side and with the superior thyroid. In its course upwards, this artery gives small twigs to the longus colli, œsophagus and trachea, and also to the scalenus anticus, and one branch, which is constant, the CERVICALIS ASCENDENS, which ascends in front of the scalenus anticus and longus colli, to the rectus anticus major, giving twigs to these muscles, and also branches which bend backwards to the muscles under the occiput.

The inferior thyroid often gives off the *supra-scapular artery*, and the *transversalis colli*, and is then a large trunk, which, sending off a leash of arteries, has been styled the *thyroid axis*.

Sometimes a *third* artery is sent to the thyroid gland, derived from the carotid, subclavian, or aortic arch; when present, it is always found ascending in front of the trachea *.

Downwards the subclavian sends off

3. The INTERNAL MAMMARY ARTERY:—This artery comes off nearly opposite to the inferior thyroid, and enters the chest, passing downwards behind the carti-

---

* *Artery of Neubauer. A. thyroidea ima.*

lage of the first rib. It then descends, gradually de-creasing in size, on the inside of the cartilages of the ribs, near the edge of the sternum, being separated from the cavity of the chest by the pleura, to which the artery is external:—at length, near the ensiform cartilage, it divides into two branches, which are distributed to the parietes of the abdomen, inosculating with the epigastric and other arteries.—The internal mammary is a large artery, and its branches are numerous; it gives off arteries to the thymus, to the anterior mediastinum, and one artery, the *Superior Phrenic* or *Comes nervi Phrenici,* which is constant, and, with two veins, accompanies the phrenic nerve to the diaphragm : branches of the mammary also pass to the intercostal spaces, and penetrate to the outside of the chest, where they communicate with the external thoracic arteries.

4. The Superior Intercostal Artery arises from the lower and back part of the subclavian, nearly at the same point as the Cervicalis Profunda, of which it is sometimes a branch. It passes backwards into the chest in front of the neck of the first rib, dividing into branches, which supply the two or three upper intercostal spaces.

Outwardly the subclavian sends off.

5. The Arteria Transversalis Colli, or *Posterior Scapular.*—This is generally a considerable artery, some-times arising from the thyroid axis. It proceeds trans-versely across the neck, somewhat deeply, crossing the scaleni muscles and nerves of the brachial plexus, or sometimes passing through the plexus. It then curves backwards, under the trapezius and levator sca-pulæ, and reaching the posterior border of the scapula, continues its course directly downwards, under the rhom-boid muscle, along the bone to its inferior angle, distri-buting numerous branches to the adjacent muscles.

Near its origin this artery gives off several branches, which ascend to the scaleni, and also a larger and more superficial branch, the SUPERFICIAL CERVICAL ARTERY, which passes backwards, and is distributed to the trapezius and splenius, and to the skin and cellular tissue at the lateral and back parts of the neck.

6. The SUPRA-SCAPULAR ARTERY, or *Dorsalis Scapulæ*, is smaller than the last-described artery; it often arises with it by a common trunk, or is a branch of the inferior thyroid. This artery also crosses the neck transversely, at its lower part, passing in front of the scaleni and behind the mastoideus, and following the line of the clavicle, above and posterior to which bone the supra-scapular is placed, in its outward course to reach the notch in the superior border of the scapula. It passes above, (or, with the supra-scapular nerve, sometimes beneath,) the ligament which converts this notch into a foramen, turns outwards along the supra-spinal fossa, supplying the supra-spinatus muscle, and then directs its course beneath the root of the acromion to the infra-spinal space, where it ramifies close to the bone under the infra-spinatus muscle.

7. The ARTERIA CERVICALIS PROFUNDA or POSTERIOR, is variable in its origin, sometimes coming off by a common trunk with the superior intercostal, or it is derived from the thyroid or vertebral arteries, or it arises from the back part of the subclavian itself, just where it is passing between the scaleni. It ascends obliquely outwards, passes between the transverse processes of the two last cervical vertebræ, and is then continued upwards on the back part of the spinal column, lying close to the bones, ramifying largely on the muscles in the posterior part of the neck, and communicating with the occipital and vertebral arteries.

COURSE OF THE SUBCLAVIAN VEIN.

The Subclavian Vein is situated anteriorly to the sub-clavian artery. This great Vein is the continued trunk of the Axillary Vein, which comes up from the axilla under the arch of the pectoralis minor, in front of, and closely applied to, its corresponding artery : with the artery, it sinks beneath the clavicle and subclavius muscle, and then passes transversely inwards, over the first rib, into the chest, crossing before the belly of the scalenus anticus, which muscle is thus interposed between the subclavian Artery and Vein.

BRANCHES. The Subclavian Vein, on each side, re-ceives behind the clavicle the external and internal Jugular Veins ; the internal jugular is of large size, and opens into the subclavian, just where it is passing into the chest ; on the left side, the long transverse venous trunk, thus formed, receives the left internal mammary, vertebral, superior intercostal,. and inferior thyroid veins, which last vein descends from a large venous plexus in front of the trachea. On the right side, the united venous trunk is short, and usually the right mammary and inferior thyroid veins join the superior cava.

The COURSE of the AXILLARY or BRACHIAL PLEXUS of nerves may also be examined. This plexus is formed by branches of the four lower cervical and first dorsal nerves, which pass between the anterior and middle sca-leni muscles into the axilla. The five nerves unite by cross branches, and form a plexus of considerable breadth extending from the side of the neck to the axilla : the up-per part of the plexus lies between the scaleni, and above and behind the subclavian artery : as the nerves descend, they emerge from behind the anterior scalenus, and pass obliquely downwards beneath the clavicle, to reach the hollow of the axilla : to this point the plexus is placed.

behind the artery and vein, but in the axilla the principal branches will be found to surround the artery with a sort of sheath.

A considerable part of the scaleni muscles may now be seen; the upper insertion of these muscles must be dissected with the muscles of the back part of the neck.

1. The SCALENUS ANTICUS—*Arises,* by four tendons, from the anterior tubercles of the transverse processes of the third, fourth, fifth, and sixth vertebræ of the neck : these unite and form a flat muscle.

Inserted, tendinous and fleshy, into the external surface and upper edge of the first rib, at a *tubercle* or eminence placed immediately before the groove for the subclavian artery.

2. The SCALENUS MEDIUS—*Arises,* tendinous, from the posterior tubercles of the five or six lower cervical transverse processes : its fleshy fibres descend to be

Inserted into the upper and outer part of the first rib, from its root, as far forward as the posterior border of the groove for the subclavian artery.

3. The SCALENUS POSTICUS—*Arises,* tendinous, from the transverse processes of the fifth and sixth vertebræ of the neck.

Inserted into the first rib, near the spine, and also into the upper and back part of the second rib.

Situation : These muscles, extending from the sides of the cervical vertebræ to the first rib, are covered before by the sterno-mastoideus and trapezius, behind by the trapezius and levator scapulæ. The anterior and middle scaleni are separated by a triangular interval, with its base on the first rib, which gives passage to the subclavian artery below, and, above, to the nerves of the brachial plexus. The two last scaleni are closely united, and are by some described as one muscle. Supernumerary fasciculi are occasionally found, separating the nerves of the axillary plexus *.

Use : To bend the neck laterally ; and, when the neck is fixed, to elevate the ribs.

SECTION IV.

DISSECTION OF THE AXILLA, OR ARM-PIT.

The Axilla is formed by two muscular folds, which bound a middle cavity. The anterior fold is formed by the pectoralis major passing from the thorax to the arm, the posterior by the latissimus dorsi coming from the back. In the intermediate cavity there is a quantity of cellular membrane and absorbent glands, covering and connecting the great vessels and nerves : below, the cavity is closed by some aponeurotic fibres passing upwards from the ribs to both borders of the axilla.

If the Pectoralis major and minor have been dissected, and reflected from their origins in the thorax, the cavity of the axilla will now be open at its anterior part : the axillary vessels are not immediately exposed, as they are concealed by a quantity of loose cellular and adipose tissue, which is continued from the interstice above the clavicle, and also by the *fascia*, which descends over the subclavius muscle to the ribs. This fascia extends outwards to the coracoid process, is very distinct, and frequently has a firm defined edge, and the great vessels are seen coming from behind it.

Clear away the cellular tissue, which is intermixed with the thoracic branches of the axillary plexus. The Axillary Vein will be found lying anterior to the artery, and passing, in the lower part of the axilla, to its inner side. It seems to be a continuation of the basilic vein, and of the two *venæ satellites,* or veins, which accompany the brachial artery. It receives branches cor-

* *Scalenus minimus* of Soemmering.

responding to the ramifications of the axillary artery. Passing upwards under the clavicle, it becomes the sub-clavian vein, and runs over the first rib, and before the anterior scalenus muscle into the thorax : just before it passes beneath the clavicle, it usually receives the Cephalic vein.

Deeper seated, and immediately behind the axillary vein, lies the AXILLARY ARTERY. It is seen coming from under the clavicle and subclavius muscle, over the border of the first rib; it lies external to the vein, or nearer the shoulder, but is overlapped by it; the nerves of the brachial plexus are close on its outer side. It next descends under the arch formed by the pectoralis minor, resting, posteriorly, on the first intercostal space, and on the upper digitation of the serratus magnus; then passing off from the chest to the arm, it becomes surrounded, in the axillary cavity, by the meshes of the Nerves, and runs along the inferior edge of the coraco-brachialis muscle. It retains the name of Axillary artery from the border of the first rib to the lower boundaries of the axilla. When it has passed the inferior edge of the latissimus dorsi, it assumes the name of the BRACHIAL ARTERY.

The branches of the Axillary artery are,

1. The EXTERNAL THORACIC ARTERIES :—these are commonly three or four in number, and come off either separately, or by common trunks, which subdivide.

(a,) The *Superior Thoracic* commonly arises along with the next artery ;—it descends obliquely between the pectoralis major and minor, to which muscles it is distributed and to the adjacent parts. Sometimes there are two or three superior thoracic arteries.

(b,) The *Humeral Thoracic*, or *Acromial Artery*, is of

M

considerable size, especially where it gives off the last branch. It arises from the fore part of the axillary artery, and comes forwards immediately above the pectoralis minor; it gives branches to the serratus magnus and pectoral muscles, and then inclining outwards towards the deltoid muscle, divides, behind the cephalic vein, into two branches; one of which ascends to the clavicle, and ramifies in the deltoid and about the capsule of the shoulder-joint, while the other branch takes the course of the cephalic vein between the deltoid and pectoralis major.

(c,) The *Inferior,* or *Long Thoracic,* called also the *External Mammary Artery,* comes off from the axillary lower down, and descends to the lateral part of the chest, along the lower border of the pectoral muscle, to which it gives branches, and also to the serratus magnus, intercostal muscles, mammary gland and integuments.

(d,) Another small branch is the *Thoracica Alaris,* which ramifies in the axilla and adjacent muscles.

2. The SUBSCAPULAR ARTERY, or *Inferior Scapular,* is the largest branch of the axillary artery, and comes off from its under and back part, opposite to the neck of the scapula. Having given off some short branches to the axilla and subscapularis muscle, it attaches itself to the inferior costa of the scapula, and after a course of two inches, splits into two branches.

(a,) The *inferior* or *internal branch* continues to descend along the inferior costa of the scapula, dividing into numerous branches to the latissimus dorsi, serratus magnus, and other muscles.

(b,) The *Superior* or *Dorsal branch* passes out of the axilla below the long head of the triceps, and above the

2

teres major, bending round the border of the scapula to the dorsum of the bone, and there ramifies chiefly below the spine in the substance of the muscles.

3. The POSTERIOR CIRCUMFLEX is a considerable artery; it arises from the back part of the axillary artery, and runs backwards close to the humerus, surrounds its neck, and is lost on the inner surface of the deltoid; it gives also twigs to the joint and neighbouring muscles.

4. The ANTERIOR CIRCUMFLEX is a much smaller artery, and is often a branch of the last: it turns round the fore part of the neck of the humerus, under the co-raco-brachialis and short head of the biceps, and is lost on the inner surface of the deltoid, giving also twigs to the capsule of the shoulder-joint.

The AXILLARY PLEXUS is seen passing from the side of the neck into the axilla, and is placed behind the great vessels. The nerves of the plexus are closely applied to the axillary artery, and as they descend beneath the arch of the pectoralis minor, they are united by cross branches, which pass in front and behind the artery, and, with the connecting cellular tissue, surround it, as with a sheath. In the lower part of the axillary space, the plexus again separates into distinct branches, which form the nerves of the arm.

From the axillary plexus, the following nerves pass off.

1. The THORACIC NERVES are usually three in number. Of these, two small nerves come off from the fore part of the plexus, and are seen passing down in front of the axillary vessels, to terminate on the greater and less pectoral muscles. The third, or *Posterior Thoracic Nerve* *, is a large branch, and the most remarkable; it

---

* The *Inferior External Respiratory Nerve* of Mr. Bell.

is seen descending in the back part of the axilla, closely applied to the thoracic wall of the space, and resting on the serratus magnus, to which it is distributed.

2. The SUPRA-SCAPULAR NERVE, or *External Scapular*, comes off from the upper edge of the plexus, and passes obliquely backwards towards the superior costa of the scapula, accompanying the supra-scapular artery, it passes through the supra-scapular notch or foramen, gives filaments to the supra-spinatus muscle, and then descends into the infra-spinal fossa, where it ramifies to the muscles.

3. The INFRA-SCAPULAR NERVES are two or three filaments, derived from the lower part of the axillary plexus, which pass backwards upon the subscapularis, supplying that muscle, the latissimus dorsi and adjacent muscles.

4. The CIRCUMFLEX, or AXILLARY NERVE lies deep; and is a large nerve; it passes from the back part of the plexus, goes backwards round the neck of the humerus, accompanying the posterior circumflex artery, and lying close upon the bone; it is distributed to the deltoid muscle and muscles on the outside of the arm.

The other nerves, which pass off from the axillary plexus, are five in number, and being distributed to the arm and fore arm, will be described in the dissection of the superior extremity: they are

5. The INTERNAL CUTANEOUS NERVE.

6. The EXTERNAL CUTANEOUS NERVE, or *Nervus Musculo-cutaneus*.

7. The MEDIAN NERVE.

8. The ULNAR NERVE.

9. The SPIRAL or RADIAL NERVE.

The *Lymphatic Glands* of the axilla are numerous, many lying between the vessels and thoracic wall of the space. Some filaments of the Dorsal or Intercostal Nerves are also seen in the back part of the axilla, coming out from the chest between the upper ribs, and passing to the arm.

SECTION V.

DISSECTION OF THE POSTERIOR MEDIASTINUM, AND OF THE NERVES AND VESSELS WHICH HAVE THEIR COURSE THROUGH THE THORAX.

FIRST trace the COURSE of the PHRENIC NERVE through the thorax.—On each side, this nerve is seen descending from the anterior scalenus, and entering the thorax between the Subclavian artery and vein : it then proceeds downwards before the root of the lung, and on the outside of the pericardium, betwixt the lateral part of that bag and the pleura, and is continued on to the diaphragm. The left phrenic nerve has a somewhat longer course than the nerve of the right side, as it turns over the pericardium, where that bag covers the apex of the heart. The phrenic nerve is accompanied by one artery and two veins; it is distributed to the diaphragm, and some filaments traverse the muscle, and ramify on its abdominal surface, communicating with the plexuses about the cœliac artery.

Now proceed to the examination of the POSTERIOR MEDIASTINUM. Saw through the ribs on each side, at the distance of three or four inches from the spine; raise either lung from its cavity, and turn it towards the opposite side of the chest; you expose the pleura reflected from the under surface of the root of the lungs to the spine and ribs. A triangular space or cavity is here formed between the two pleuræ and the bodies of the dorsal vertebræ. This is named the *Cavity of the Pos-*

M 3

terior Mediastinum. It contains many important parts, viz. the trachea, œsophagus, aorta, nerves of the eighth pair, vena azygos, and thoracic duct, with lymphatic glands and some cellular tissue.

This cavity may be laid open for examination, either on the right or on the left side. Fold back the lungs to the *left* side of the chest; divide the pleura longitudinally, where it passes from the right side of the spinal column to the root of the right lung. You thus lay open the *posterior mediastinal cavity* on the *right side;* remove cautiously the cellular tissue, and observe the relative situation.

But first let us attend to the course of the INTERCOSTAL or GREAT SYMPATHETIC NERVE, which is seen running by the side of the spine. The two great sympathetic nerves are not properly contained within the posterior mediastinum, but descend along the sides of the spinal column. The nerve of each side, where it enters the thorax, is situated deep behind the great vessels. The *inferior cervical ganglion* has been already noticed in the neck; it is now apparent in the thorax, placed between the transverse process of the seventh cervical vertebra and the neck of the first rib; it sends off filaments, which encircle the subclavian artery and some of its branches. From the inferior cervical ganglion, the sympathetic nerve descends along the thorax: it lies upon the heads of the ribs, where they are articulated with the vertebræ, forming a ganglion in each intercostal space, and receiving twigs from each of the dorsal nerves, just after their exit from the vertebral canal. It lies behind the pleura, but is seen through it; it passes into the abdomen by the side of the spine, running through the fibres of the small muscle of the diaphragm.

BRANCHES of the SYMPATHETIC NERVE in the THORAX.

Filaments are sent off from the thoracic ganglia, to the mediastinum and descending aorta, and to join the pulmonary plexus of the Nervus Vagus: but the Splanchnic Nerves are the most worthy of remark, giving rise to the great abdominal plexuses. The greater SPLANCHNIC NERVE, or *Anterior Intercostal*, is formed by three to six filaments, which come off from the thoracic ganglia of the sympathetic between the sixth and tenth dorsal vertebræ: these filaments passing forwards on the bodies of the vertebræ, behind the pleura, unite to form the *splanchnic nerve*, which appears as a single nervous chord about the eleventh dorsal vertebra, and may be traced entering the abdomen between the fibres of the lesser muscle of the diaphragm. The *Lesser Splanchnic Nerve* is formed by two filaments of the tenth and eleventh thoracic ganglia; these unite to form a small nerve, which pierces the diaphragm separately.

The right sympathetic nerve lies under the pleura by the right side of the spine. Still nearer the middle of the spine, you see the VENA AZYGOS. In dissecting, you find it situated betwixt the right sympathetic nerve and the aorta; it begins below from ramifications of the lumbar veins, and it enters the chest by piercing the small muscle of the diaphragm, or by passing between the crura with the aorta. This vein ascends, in the thorax, along the spine; it is placed to the right of the aorta, and on the right side of the spinal column, receiving *veins* from each of the *intercostal spaces* of the right side, and also some *œsophageal veins*, and a *bronchial vein* from the right lung: and, about the middle of the back, it receives a considerable trunk, the *Azygos Sinistra*, which comes from under the aorta, and returns the blood from the left side of the thorax. At the fourth dorsal vertebra, the vena azygos leaves the spine; it makes a curve forward and upward, passing over the right bronchus, and empties its blood into the back part of the vena cava superior, immediately before that vein enters the pericardium.

Descending through the posterior mediastinum, will be also found the AORTA. The great artery, having formed its arch, comes in contact with the third dorsal vertebra, and is now called the *Descending* or *Thoracic Aorta*. It descends along the bodies of the dorsal vertebræ, rather on their left side; it lies behind the œsophagus, and passes betwixt the crura of the diaphragm into the abdomen.

BRANCHES OF THE AORTA IN THE THORAX.

1. The BRONCHIAL ARTERIES are two, sometimes three or four, small twigs of the aorta, which pass to the lungs on each side.

2. Small arteries pass forwards from the aorta to the œsophagus, named *A. Œsophagæ* :—others run to the posterior mediastinum.

3. The INFERIOR or AORTIC INTERCOSTALS are eight or nine in number on each side of the thorax; they come off separately from the side or back part of the aorta, and seem to tie that great artery to the spine. Each intercostal artery passes immediately into the interval betwixt two ribs, and there subdivides into

(1.) A *posterior* branch, which perforates between the heads of the ribs, to the muscles of the back : this branch also gives twigs which enter the spinal canal.

(2.) The continued trunk of the artery runs forwards in the groove in the inferior edge of the rib, between the two layers of intercostal muscles, accompanied by a nerve and one or two veins. It gives branches to the intercostal muscles and parietes of the chest, and, reaching the anterior part of the thorax, is lost in the muscles, inosculating with the internal mammary and thoracic arteries.

The intercostal arteries of the right side are the longest, crossing over the bodies of the vertebræ. These aortic intercostals also communicate with the superior intercostals, which come off by one common trunk, from the subclavian.

The dissector also finds in the posterior Mediastinum the THORACIC DUCT. He must look for it behind the œsophagus, betwixt the vena azygos and aorta; it is of the size of a crow-quill; but is collapsed, and appears like cellular membrane condensed, and can only be distinguished when inflated or injected; it was seen in the abdomen close to the aorta, passing into the thorax between the crura of the diaphragm. It ascends along the posterior mediastinum, and, about the sixth dorsal vertebra, bends obliquely, to the left side, behind the œsophagus, and then continues its course behind the great arch of the aorta, and on the inner side of the left subclavian artery, lying on the longus colli muscle. Having ascended as high as the seventh cervical vertebra, it curves downwards and inwards, passes behind the inferior thyroid artery and left internal jugular vein, and enters the left subclavian vein, at the point where that vein is joined by the left internal jugular. Slit open the vein, and you find the opening of the duct furnished with two membranous valves. The duct is commonly tortuous in its course, and contracts and again enlarges, and it frequently splits into several branches, which re-unite. The absorbents of the right superior extremity, and of the right side of the head and thorax, usually form *another trunk*, which enters the point of union between the right subclavian vein and right internal jugular.

The ŒSOPHAGUS is also situated between the layers of the posterior mediastinum. It lies immediately before the aorta, at first towards its right side, but crossing, inferiorly, to the left. It is seen descending from the neck behind the trachea, at first lying on the spine, then passing in front of the descending aorta: it crosses behind the left bronchus, and is then placed immediately beneath the pericardium and base of the heart. It passes through the oval opening or fissure in the diaphragm, and immediately expands into the stomach.

The TRACHEA is also seen, behind the arch of the aorta and great vessels proceeding from the heart, descending from the neck into the upper and posterior part of the chest. It enters the thorax between the two pleuræ, and, opposite to the second or third dorsal vertebræ, bifurcates into the two *Bronchi*, one of which passes toward the right, the other toward the left, to enter the lung of each side, and, subdividing, to ramify in its substance. About the bifurcation of the trachea, numerous Lymphatic glands are observed, the *Bronchial Glands*: in the child, reddish, but in the adult of a brown or black colour; they are placed in front of the bifurcation of the trachea, around the bronchial tubes, and are even found in the substance of the lung.

COURSE OF THE PAR VAGUM or NERVUS VAGUS IN THE THORAX. From the neck, the nervus vagus is seen descending into the chest behind the subclavian vein: the nerve of the right side crosses before the right subsubclavian artery; on the left side, the nervus vagus descends longitudinally in front of the left subclavian artery, and passes before the arch of the aorta. Each nerve immediately sends off a large branch, the *Recurrent* or *Inferior Laryngeal Nerve*, into the neck: on the right side, this branch is seen twisting round under

the subclavian artery; on the left side, under the arch
of the aorta; it ascends behind the carotid, and lodges
itself between the trachea and œsophagus, in which situation it has been dissected in the neck.

The Nervus Vagus, having given off the Recurrent,
descends by the side of the trachea, and behind the root
of the lungs. It gives off five or six filaments, which
run down in front and behind the trachea, then, opposite
the bronchus of each side, the Vagus augments much in
volume,—its filaments separate and form a kind of net-
work, intermixed with cellular tissue and many vessels,
and in this way there is formed, behind the lung, a very
complicated plexus, termed the *Pulmonary Plexus,*
which receives some branches from the great sympathetic,
and from which numerous filaments go to the lung, accompanying the divisions of the bronchial tube.

Below this pulmonary plexus, the filaments of the
Nervus Vagus re-unite to form a nervous chord, which
attaches itself to the œsophagus, the left nerve running
down on the fore part of the tube, the right nerve on the
back part. These two nervous chords are connected by
frequent anastomoses, forming the *Œsophageal Plexus,*
and then continuing to descend, pass upon the œsophagus through the diaphragm, to ramify on both surfaces
of the stomach.

The twelve DORSAL NERVES are also seen in this dissection, emerging from the spinal canal, between the
bodies of the vertebræ, and dividing into their *anterior*
and *posterior* branches, of which the former, passing
into the intercostal space, soon incline to the lower border of the rib above, and are found accompanying the
intercostal arteries.

SECTION VI.

DISSECTION OF THE HEART, WHEN REMOVED FROM THE BODY, AND OF THE STRUCTURE OF THE LUNGS.

THE *Structure* of the heart is muscular. It has, 1. An external smooth tunic or coat, which is a reflection of the internal serous lamina of the pericardium. 2. Its internal surface is lined by a fine transparent membrane, which is strongly adherent, smooth and lubricated, and continuous with the lining membranes of both arteries and veins. 3. Between these two membranous tunics is the muscular substance or third coat, which varies in thickness at different parts. In the right side of the heart, we always meet with a considerable quantity of coagulated blood. In the left side, there is much less. External to the muscular tunic, in adults and the aged, a considerable quantity of fat is observed.

First examine the *right side* of the heart.

1. Slit open, with the scissors, the two VENÆ CAVÆ on their fore part; the inner surface of these veins, and of the RIGHT AURICLE, will be seen lined by a smooth membrane; and in the auricle, the *musculi pectinati,* or bundles of muscular fibres, will be seen projecting. At the point of union between the two cavæ, there is a projection formed by the thickening of the muscular coat, the *Tuberculum Loweri.* The *Septum Auricularum* is seen separating the right from the left auricle :—observe that it is thin, that in it there is an oval depression, named *Fossa Ovalis.* Round this fossa the fibres are thicker, forming the *Annulus Ovalis;* this is the remains of the *Foramen Ovale* of the fœtus ; and, in many adult subjects, a probe may be passed through the superior part of the fossa obliquely into the left auricle. The

Valvula Nobilis, or *Eustachian Valve,* is a membrane-like duplicature of the inner coat of the auricle, observed where the vena cava inferior is continued into the auricle, and stretching from that vein towards the opening into the right ventricle. This valve is sometimes found reticulated. Behind this valve, is the orifice of the *Coronary Vein,* with its small *valve.*

The *Foramina Thebesii* are minute orifices of veins, which open into all the cavities of the heart; they are most numerous, however, in the right auricle.

The *Ostium Venosum,* or opening of the right auricle into the right ventricle, is somewhat oval; it has a valve, which projects into the right ventricle.

2. The RIGHT VENTRICLE may now be opened by an incision, carried from the root of the pulmonary artery down to the apex of the heart. This incision should be made with care, lest the parts on the inside of the ventricle be destroyed by it; it should pass along the right side of the *Septum Ventriculorum,* the situation of which is marked out by large branches of the coronary artery and vein. A small opening should first be made, into which one blade of the scissors can be introduced: the incision may be continued through the apex of the heart; or a flap may be made by another cut, passing from the beginning of the first along the margin of the right auricle.—In this ventricle, observe the projecting bundles of muscular fibres; the *Tricuspid Valve,* arising from the margin of the ostium venosum, or *auriculo-ventricular* opening, and projecting into the right ventricle. This valve forms a complete circle at its base, but has its edge divided into three parts, which are attached by tendinous filaments, named *Chordæ Tendineæ,* to the *Carneæ Columnæ,* or muscular bundles of the ventricle.

Slit up the PULMONARY ARTERY; observe how it

arises from the back part of the right ventricle, how smooth the inside of the ventricle becomes as it approaches the entrance of the artery, or *ostium arteriosum.* Observe the three *Semilunar* or *Sigmoid Valves.* Their bases arise from the artery, their loose edges project into its cavity, and in the middle of the loose edge of each valve is seen a small white body, termed *Corpus Sesamoideum.* The artery is seen bifurcating into the right and left pulmonary arteries, and, just before its bifurcation, sending off to the aorta the *ductus arteriosus,* which in the adult is a ligament.

We now proceed to examine the left side of the heart. 3. The Left Auricle has four *pulmonary veins* opening into its cavity, which may be exposed by slitting up two of these veins. Observe that its walls are thicker than those of the right auricle. The *septum auricularum,* with the *fossa ovalis,* is here seen less distinctly than on the right side. Observe also the *ostium venosum,* opening into the left ventricle, and giving attachment to the *Valvula Mitralis.*

4. The Left Ventricle may be opened in the same manner as the right, by an incision carefully made on the left side of the septum or partition of the ventricles, and continued round the upper part of the ventricle, under the auricle. Observe the great thickness of the muscular coat; the *Valvula Mitralis,* descending from the auriculo-ventricular opening, and forming two projections, which are attached by the *chordæ tendineæ* to the *fleshy columns* of this ventricle.

Slit up the Aorta; it has three *semilunar valves;* which resemble those of the pulmonary artery;—behind these valves the artery bulges out, forming the *Sinuses* of the aorta. Above two of the valves lie the orifices of the two *Coronary Arteries,* which supply the heart.

The *Right* or *Posterior Coronary Artery* runs transversely in the depression, which separates the right auricle from the right ventricle, turns round the base of the heart, and is continued along the posterior flattened surface, in the groove between the two ventricles. The *Left* or *Anterior Coronary Artery* is the smallest; it passes down between the pulmonary artery and the left auricle, and arriving at the groove between the ventricles on the anterior surface of the heart, descends along it to the apex. The *Veins* accompany the arteries, and the greater number unite to form the *Coronary Vein*.

STRUCTURE OF THE LUNGS. Observe their appearance and texture; spongy, cellular, and expansible; invested externally by the serous pleura, lined in their interior by a prolongation of mucous membrane; of a pale or greyish white colour, with small dark spots; receiving at their roots the bronchi and pulmonary vessels. The substance or *parenchyma* of the lung is very complex, apparently consisting of minute lobules, which are composed of the ultimate ramifications of the bronchial tubes, pulmonary arteries and veins, with nerves and lymphatics, united and supported by a fine cellular tissue.

Examine the TRACHEA. Shape, cylindrical, but flattened behind; formed of sixteen to twenty incomplete *rings* of *fibro-cartilage*: these rings are placed horizontally below one another, and are united by an elastic *ligamentous* substance, which also fills up the posterior interval: within this ligamentous layer, some *transverse fibres*, of a muscular appearance, are observable, uniting the extremities of the rings of cartilage: more internally, is the *mucous membrane*, lining the trachea, continued from the larynx, of a reddish colour, and having numerous excretory orifices of mucous follicles or glands. In, the bronchial ramifications, the cartilaginous structure gradually disappears.

The Œsophagus consists of two tunics or coats. 1. The *Muscular*, which is composed of two layers of fibres; the external are longitudinal, and, diverging, ultimately pass upon the stomach: the internal are circular. 2. The *Inner* or *Mucous* coat is a soft and whitish membrane, continued from the pharynx, but contrasting, in its pale colour, with the redness of the pharyngeal lining: it is thrown into longitudinal plicæ, and has numerous mucous follicles; below, it joins the mucous lining of the stomach, but the two membranes are not continuous, the œsophageal lining terminating at the cardia by a fringed tuberculated line.

CHAPTER IX.

DISSECTION OF THE FACE.

SECTION I.

OF THE MUSCLES.

UNDER the integuments of the face, there is always a considerable quantity of adipose tissue; many of the muscles are very slender, their fibres pale, and, lying embedded in this fat, they require careful dissection. The whole side of the face is also supplied with numerous ramifications of the facial nerve, or portio dura of the seventh pair. These nervous twigs are generally removed with the integuments.

Twelve pairs of muscles, and one single muscle, are described in this dissection.

Make a circular incision round the orbit; this will expose a muscle surrounding the bony circle, covering

the eyelids, and connected with the cheek and forehead,

1. The ORBICULARIS PALPEBRARUM.—It *arises*, by
fleshy fibres, from a short round tendon at the inner
angle of the eye, from the front edge of the groove of the
lachrymal sac, and from the neighbouring part of the
base of the orbit : the fibres pass downwards and outwards, spreading over the under eyelid and upper part
of the cheek; they surround the outer angle of the orbit,
and then run inwards over the superciliary ridge, covering also the upper eyelid, to be

Inserted into the internal angular process of the frontal bone, and into the same short tendon, which is fixed
to the nasal process of the superior maxillary bone, and
which serves to give attachment to the cartilages of the
tarsi, and to the fibres of this muscle.

Situation: This muscle is intermixed, at its upper
part, with the occipito-frontalis; below, it is loosely connected with the fat and muscles of the cheek; its posterior surface is applied to the bony circle of the orbit,
and to the cartilages of the eyelids, and, at the inner
canthus, it covers the lachrymal sac.

That part of the orbicularis, which covers the lids, is
sometimes described as a distinct muscle, under the
name of *Ciliaris.*

Use: To shut the eyelids, and to compress the eye-
ball and lachrymal sac.

Separate the fibres of the occipito-frontalis, near the
inner corner of the orbit, from the orbicularis palpebrarum : you expose a small muscle.

2. The CORRUGATOR SUPERCILII—*Arises*, fleshy, from
the internal angular process of the os frontis; it runs
outwards and a little upwards, to be

Inserted into the inferior fleshy part of the occipito-frontalis muscle, extending outwards as far as the middle of the superciliary ridge.

Situation: This muscle is situated in the thick inner part of the brow, concealed by the occipito-frontalis, and applied to the upper and inner part of the orbicularis palpebrarum ; its anterior surface is covered by these two muscles : its posterior surface rests on the frontal bone, and on the supra-orbitar artery and nerve.

Use: To draw the eyebrow and skin of the forehead downwards, into vertical wrinkles.

3. The Compressor Naris—*Arises,* narrow, from the outer part of the ala nasi, and neighbouring part of the superior maxilla. From this origin, a number of thin diverging fibres run up obliquely, along the cartilage of the ala nasi towards the dorsum, where the muscle joins its fellow, and is

Inserted, slightly, into the lower part of the os nasi and nasal process of the superior maxilla.

Situation: This is a thin triangular muscle, placed transversely upon the ala nasi ; its origin is connected with the levator labii superioris alæque nasi ; and its upper part with the descending slip of the occipito-frontalis, or *pyramidalis nasi.*

Use: It is said to compress the ala nasi, but it rather serves to dilate the anterior nares.

We now proceed to dissect the muscles of the mouth.

An incision round the lips exposes the orbicularis oris, into which all the other muscles pass from the side of the nose, cheek, and lower jaw. The scalpel is to be carried in the direction of their fibres, reflecting the integuments cautiously, and the cheek being distended by horse-hair or a sponge :—these fibres become more visible by exposure to the air after the dissection.

4. Levator Labii Superioris Alæque Nasi—*Arises*

by two distinct origins;—the first, from the nasal process of the superior maxilla, where it joins the os frontis at the inner canthus of the eye: it descends along the nasal process, and is *inserted*, by two slips, into the outer part of the ala nasi, and into the upper lip.

The second *arises*, broad and fleshy, from the external orbitar process of the superior maxilla, immediately above the foramen infra-orbitarium; runs down, becoming narrower, and is *inserted* into the upper lip and orbicularis oris.

Situation : This muscle is superficial. The first portion is sometimes called Levator Labii Superioris Alæque Nasi, and the second Levator Labii Superioris Proprius. Their origins are partly covered by the lower border of the orbicularis palpebrarum. They descend more outwardly than the ala nasi.

Use : To draw the upper lip and ala nasi upwards and outwards.

The *infra-orbitar vessels* and *nerve* are seen emerging from the infra-orbitar foramen under this muscle, and passing down over the levator anguli oris.

5. ZYGOMATICUS MINOR—*Arises* from the upper prominent part of the os malæ, and, descending obliquely downwards and forwards, is

Inserted into the upper lip near the corner of the mouth.

Situation : This slender muscle has its origin covered by the orbicularis palpebrarum; it descends between the levator labii superioris and zygomaticus major. Frequently the orbicularis palpebrarum sends down a slip of fibres to the upper lip; this slip runs between the zygomaticus minor and levator labii superioris, or supplies the place of the zygomaticus minor, which is not always found.

Use : To draw the corner of the mouth upwards.

6. ZYGOMATICUS MAJOR—*Arises,* fleshy, from the os malæ, near the zygomatic suture.

Inserted into the angle of the mouth, appearing to be lost in the depressor anguli oris, and orbicularis oris.

Situation : Its origin is partially covered by the orbicularis palpebrarum. This long rounded muscle crosses obliquely the side of the face, more outwardly than the zygomaticus minor : its posterior surface covers the os malæ, portions of the buccinator and masseter muscles, and more or less of fat.

Use: The same as the last.

7. The LEVATOR ANGULI ORIS, or *Musculus Caninus* —*Arises,* thin and fleshy, from a depression * of the superior maxilla, just below the foramen infra-orbitarium ; it descends obliquely.

Inserted, narrower, into the angle of the mouth.

Situation : This is a deep-seated muscle, resting on the maxilla, buccinator, and mucous membrane of the mouth : it lies more outwardly than the levator labii superioris, and descends to the angle of the mouth behind the outer part of that muscle and the zygomaticus minor, being in part concealed by those muscles, and by the infra-orbitar nerve and vessels. At its insertion, it is particularly connected with the depressor anguli oris.

Use: To draw upwards the corner of the mouth, and upper lip.

8. The DEPRESSOR ANGULI ORIS, or *Triangularis Oris*—*Arises,* broad and fleshy, from the lower edge of the inferior maxilla, at the side of the chin, and gradually becoming narrower, is

Inserted into the angle of the mouth.

* *Fossa Canina.*

Situation : This is a flat triangular muscle, attached to the jaw from the edge of the masseter to the foramen mentale ; it is firmly connected with the skin and with the platysma myoides ; at its insertion, it is blended with the zygomaticus major and levator anguli oris.

Use : To pull down the corner of the mouth, and elongate it transversely.

9. The DEPRESSOR LABII INFERIORIS—*Arises*, fleshy and broad, from the side of the lower jaw, a little above its lower edge ; it runs obliquely upwards and inwards, in the under lip, and is

Inserted into the edge of the lip, where it decussates with its fellow, and is closely intermixed with the orbicularis oris.

Situation : This quadrilateral muscle arises from the lower jaw beneath the depressor anguli oris, and is in part covered by it; it forms the thick part of the chin, and has its fibres interwoven with fat ; it adheres firmly to the skin : it covers the vessels and nerve which emerge from the foramen mentale.

Use : To pull the under lip downwards, and invert it.

10. The BUCCINATOR—*Arises*, chiefly fleshy, from the lower jaw, as far back as the root of the coronoid process ; from the upper jaw, as far back as the pterygoid process of the sphenoid bone ; and, in the interval, from a tendinous band *, which extends from the apex of the internal pterygoid plate to the root of the coronoid process : it then continues to arise from the alveolar processes of both jaws, as far forwards as the dentes cuspidati. The fibres run horizontally forward, converging slightly, and are

_ * *Intermaxillary* or *pterygo-maxillary ligament :* it gives attachment posteriorly to the superior constrictor of the pharynx.

Inserted into the angle of the mouth, intermixing with the orbicularis.

Situation : This is a broad, flat, muscle, extended between the jaws, and lined internally by the mucous membrane of the mouth.   It lies deep, a quantity of soft fat being always found between its fibres and the other muscles and integuments, particularly in the middle of the cheek, where it forms globular masses.   The buccinator is partly concealed by the masseter, and by the muscles which pass to the angle of the mouth ; it is inserted behind these muscles.   In the cheek it is connected with the platysma myoides, and its fibres are pierced by the parotid duct.

Use : To draw the angle of the mouth backwards, and to contract its cavity, by pressing the cheek inwards.

The single muscle is the

ORBICULARIS ORIS.—It consists of two planes of semicircular fibres, which decussate at the angles of the mouth.   These fibres are formed chiefly by the muscles which are inserted into the lips ; they surround the mouth.   The superior portion runs along the upper lip, the inferior portion along the under lip.

Situation : This muscle is placed between the skin and mucous membrane, forming the fleshy substance of both lips : to the skin it is closely adherent.   It is connected and intermixed with the insertions of all the preceding muscles of the face.   A fleshy slip sent up, on each side, to the septum nasi, has been termed by Albinus Nasalis Labii Superioris.

Use : To shut the mouth by contracting and drawing both lips together.   It is the antagonist of all the other muscles of the lips.

11. DEPRESSOR LABII SUPERIORIS ALÆQUE NASI— *Arises*, thin and fleshy, from the superior maxilla near

the sockets of the cuspidatus and incisor teeth : thence it runs up under part of the levator labii superioris alæque nasi.

Inserted into the upper lip and root of the ala nasi.

Situation: This small muscular fasciculus, placed close to its fellow, is concealed by the orbicularis oris and levator labii. It may be discovered by inverting the upper lip, and dissecting through the mucous membrane on the side of the frenum, which connects the lip to the gums.

Use: To depress the upper lip and ala nasi.

12. The LEVATOR LABII INFERIORIS—*Arises* from the lower jaw, at the root of the alveolus of the lateral incisor.

Inserted into the under lip and skin of the chin.

Situation: These two small muscles are found by the side of the frenum of the lower lip. They lie under the depressor labii inferioris.

Use: To raise the under lip and skin of the chin.

On the side of the face we observe two strong muscles, and two other muscles are concealed by the angle of the inferior maxilla.

1. The MASSETER is divided into two portions, which decussate one another.

The Anterior Portion *arises,* chiefly tendinous, from the superior maxillary bone, where it joins the os-malæ; from the lower edge of the os malæ, and from its zygomatic process. The strong fibres run obliquely downwards and backwards, and are *inserted* into the outer surface of the side of the lower jaw, extending as far back as its angle.

The Posterior portion is much smaller, lying beneath the anterior, but distinctly separated from it. It *arises,* principally fleshy, from the inferior surface of the os malæ, and of the whole of the zygomatic process, as far back as the tubercle before the socket for the condyle of the lower jaw. The fibres slant forwards, and are *inserted*, tendinous, into the outer surface of the coronoid process of the lower jaw.

Situation : This strong muscle covers the ramus of the lower jaw; its two layers are distinct, and might be considered as separate muscles, the anterior concealing almost the whole of the posterior portion. The muscle itself is, in great part, superficial. Below, it is covered by the platysma myoides : and, above, partially by the origin of the zygomaticus major. The parotid gland also laps over the posterior edge of the masseter, and the duct crosses it, to perforate the buccinator.

Use : When both portions act, to raise the lower jaw ; or acting separately, to carry the jaw forwards or backwards.

The next muscle is covered by a strong fascia, which must be slit up to expose the fibres.

2. The TEMPORALIS—*Arises*, fleshy, from a semicircular ridge in the lower and lateral part of the parietal bone, from all the squamous portion of the temporal bone, from the external angular process of the os frontis, from the temporal process of the sphenoid bone, and from the inner surface of the aponeurosis, which covers the muscle. It fills up the temporal fossa, its fibres arising from the periosteum of the fossa. From these different origins the fibres converge, and descend under the zygoma formed by the processes of the temporal and malar bones.

Inserted, by a strong tendon, into the upper part of the coronoid process of the lower jaw, to which it adheres on every side, but more particularly on its forepart, where the insertion is continued down to near the last dens molaris. .

Situation: This muscle is of a semicircular shape. The strong *fascia* covering it adheres to the bones which give origin to the upper part of the muscle, and, then descending over it, is inserted into the zygoma, and adjoining part of the os malæ and os frontis. The temporalis, at its origin, lies under the expanded tendon of the occipito-frontalis, and under the small muscles which move the external ear. Its insertion is concealed by the zygoma and by the masseter: so that, to expose it, the masseter must be cut away.

Use: To pull the lower jaw upwards.

To expose the following muscles, we must remove the muscles of the cheek and jaw;—the masseter and insertion of the temporalis must be taken away, and the coronoid process of the inferior maxilla removed by a saw.

3. The PTERYGOIDEUS EXTERNUS—*Arises,* broad and fleshy, from the outer side of the external plate of the pterygoid process of the sphenoid bone, and adjoining part of the tuberosity of the os maxillare; and from the root of the temporal process of the sphenoid bone. The fibres pass backwards and outwards, converging, to be

Inserted into a depression in the neck of the condyle of the lower jaw, and into the anterior and inner part of the inter-articular cartilage.

Situation: This short, triangular, muscle passes almost transversely from the skull to its insertion. It is concealed by the muscles of the face and neck, and by the ascending processes of the lower jaw. Its double origin often gives passage to the internal maxillary artery.

Use: When this pair of muscles act together, they bring the jaw horizontally forwards. When they act singly, the jaw is moved forwards, and to the opposite side.

N

4. The PTERYGOIDEUS INTERNUS—*Arises*, tendinous
and fleshy, from the whole pterygoid fossa, or space between the plates of the pterygoid process of the sphenoid
bone, but particularly from the inner surface of the external plate, and from the pterygoid process of the os
palati between these plates.

Inserted, by tendinous and fleshy fibres, into the inside of the angle of the lower jaw, and into the ridged
surface immediately above.

Situation: This muscle inclines outwards and backwards to its insertion, and is applied to the inside of the
ramus of the jaw, somewhat as the masseter is externally.
To expose it, the jaw must be removed from its articulating cavity, and then pulled forwards, and toward the
opposite side ; or it may be sawn across at its symphysis,
and the other half removed. It is larger than the pterygoideus externus ; and betwixt the two muscles there is
a considerable quantity of cellular tissue, and the *Dental*
and *Lingual* branches of the inferior maxillary nerve.
The circumflexus palati is applied to the inner surface
of this muscle, and along its posterior edge we observe
the Internal Lateral ligament of the lower jaw.

Use: To draw the jaw upwards, and obliquely towards the opposite side.

On the side of the face is situated the largest of the
salivary glands, the PAROTID GLAND, a large white mass,
irregularly oblong and protuberant, filling up all the
space from the zygomatic arch downwards to below the
angle of the jaw, and laterally from the ascending ramus
of the jaw to the root of the ear and mastoid process ; it
also covers a portion of the masseter muscle. The substance of the gland is extended deeply into the hollow
between the ramus of the jaw and the mastoid process,
adhering firmly to the surrounding parts, covering the

great vessels and nerves at the base of the skull, and having the external carotid artery and facial nerve embedded within its mass: on the surface of the parotid, one or two absorbent glands are generally observed. It is covered below by some of the fibres of the platysma myoides; it has no capsule, but it is bound down by a continuation of the superficial cervical fascia.

From the anterior and upper part of the Parotid, a white canal or *excretory duct* is sent off, termed *Steno's Duct.* This duct passes horizontally forwards over the masseter, about half an inch below the zygoma, and pierces the buccinator from without inwards, opening into the mouth opposite the first dens molaris of the upper jaw. There is sometimes a small separate portion or process of the gland, named the *Socia Paroditis,* situated on the masseter, above the duct.

The Parotid, like the other salivary glands, is of a greyish white colour, of a firm and resisting texture, and composed of granulated bodies united into irregular lobules and lobes. The Duct is formed of a firm, thick, and resisting substance, and is lined within by a fine mucous membrane, prolonged from that of the mouth.

SECTION II.

OF THE VESSELS AND NERVES OF THE FACE.

I. ARTERIES.

THE EXTERNAL MAXILLARY ARTERY, the third branch of the carotid, comes from the neck over the lower jaw, close to the anterior edge of the masseter: in the living subject, it is here felt distinctly pulsating, and resting on a perceptible depression of the bone. It then takes the name of FACIAL ARTERY, and runs obliquely upwards, towards the angle of the mouth, between the depressor anguli oris and masseter, being in general much contorted, and covered only by the skin and a few fibres of the platysma myoides. Near the corner of the mouth, it sinks under the union of the two muscles, levator and depressor anguli oris, and then ascends, under the name of *Ramus Nasalis* or *Angularis,* by the side of the nose, to the inner angle of the eye, where it terminates, inosculating freely with the infra-orbitar artery, and nasal and frontal branches of the ophthalmic artery.

In this course, the Facial Artery is separated from the skin by more or less of adipose substance, and gives off in succession several branches. (1.) The *Inferior Labial Artery* to the muscles and integuments of the lower lip, and *masseteric* branches outwards. (2.) The *Inferior Coronary Artery,* a larger and tortuous branch, comes off from the Facial at some distance below the commissure of the lips, and passes under the depressor anguli oris to the under lip. It runs along the loose border of the lip, close to the mucous membrane, and beneath the fibres of the orbicularis oris, meeting the artery of the opposite side. (3.) The *Superior Coronary Artery* comes off very near the commissure of the lips, and runs tortuously to the upper lip, sending twigs upwards to the tip of the nose and septum. These two coronary arteries of the lips often arise by a common trunk, which subdivides. (4.) *External Nasal Arteries,* one or more small arteries, ramifying on the side of the nose; and *Buccal* arteries to the cheek.

The EXTERNAL CAROTID is found ascending behind the parotid gland, in the space between the jaw and meatus auditorius: it is at first deeply seated, in the substance of the parotid, but, as it ascends, it advances nearer its surface. It gives small arteries to the gland,

and, opposite the neck of the lower jaw, terminates by dividing into the TEMPORAL and INTERNAL MAXILLARY ARTERIES.

The TEMPORAL is the smaller and more superficial of the two terminating branches; it continues in the direction of the carotid trunk, perforating the parotid at its upper part, and ascending over the zygoma, immediately before the ear. It then becomes subcutaneous, and passes upwards upon the temporal fascia, where it divides into two branches, an *Anterior* and *Posterior*. Before this division it gives off, (1.) *A. Transversalis Faciei;*— this is a considerable artery, sometimes arising from the carotid trunk. It comes off within the substance of the parotid gland, and, piercing its anterior edge, runs transversely over the masseter muscle, in the same line with the parotid duct, but nearer to the zygoma; it ramifies on the side of the face. (2.) *Anterior Auricular Arteries* to the external ear. (3.) The *Middle Temporal Artery* passes off near the zygoma, immediately penetrates the aponeurosis of the temporal muscle, and ramifies within the muscle, inosculating with the deep temporal branches of the internal maxillary.

The *Anterior terminating branch* of the temporal artery ascends tortuously towards the forehead, and is distributed to the muscles and integuments, anastomosing with the branches of the ophthalmic artery. The *Posterior branch* passes upwards in the same contorted manner over the parietal bone, where its branches spread widely, uniting with the posterior auricular and occipital arteries.

The INTERNAL MAXILLARY ARTERY is a larger branch than the temporal, and immediately plunges inwards behind the ascending condyle of the lower jaw; it directs its course towards the bottom of the orbit of the eye. To follow this artery, the Parotid gland, with the angle and ascending processes of the jaw, must be removed; or the coronoid process may be cut off near its root, the jaw divided at its symphysis, and then drawn outwards. The Artery is seen running inwards between the pterygoid muscles; it then ascends in a very tortuous manner, glides between the two origins of the external pterygoid, and approaching the floor of the orbit, turns horizontally into the spheno-maxillary fossa, where it is surrounded by fatty tissue, and sends off its terminating branches.

The branches of the Internal Maxillary, are, (1.) *A. Meningea Media*, or *Spheno-spinalis*, a large artery, which comes off behind the neck of the jaw, passes through the spinous hole of the sphenoid bone into the cranium, and ramifies on the dura mater. (2.) The *Inferior Maxillary*, or **Dental Artery**, descends on the inside of the ramus of the jaw, and enters, with the dental nerve, the foramen at the root of the ascending processes; it then passes along the dental canal of the lower jaw, supplying the teeth and sockets, and emerges by the foramen mentale on the chin. (3.) *A. Pterygoideæ*, and *A. Temporales Profundæ*, to the pterygoid and temporal muscles. (4.) *A. Massetericæ* and *A. Buccales* to the masseter muscle and cheek. (5.) *A. Alveolaris*, or *Superior Dental Artery*, runs tortuously along the tuber maxillare, giving branches to the teeth of the upper jaw and to the bone itself. (6.) The *Superior Pharyngeal branch*. (7.) The **Descending** *Palatine branch*, through the posterior palatine canal to the muscles and mucous membrane of the palate. (8.) The *Spheno-Palatine* or *Nasal Artery* passes through the spheno-palatine foramen to the nose. (9.) The continued trunk of the internal maxillary enters the posterior opening of the infra-orbitar canal, with its accompanying nerve, and emerges by the foramen infra-orbitarium on the face. Here it is termed the *infra-orbitar artery* ; it lies under the levator labii superioris proprius, and is distributed to the cheek and side of the nose.

The *Supra-orbitar* artery is also seen in the dissection of the forehead : it is a branch of the ophthalmic artery, coming out from the orbit with the frontal nerve, by the superciliary notch or foramen, to be distributed to the forehead. Nearer the nose, we also see the *Nasal* and *Frontal Arteries*, which are the terminating branches of the ophthalmic artery, passing out of the orbit at the inner angle of the eye, at which point a remarkable anastomosis of arteries will be observed, if the face be injected.

II. VEINS.

The Veins of the face are numerous, and pass into the external and internal jugular veins ; their general arrangement corresponds with that of the arteries.

A considerable vein may be traced from the forehead and inner angle of the eye ; it begins by several branches from the top of the head, which uniting form the *Frontal Vein;* this vein descends over the middle of the forehead, and arriving at the side of the nose, takes the name of the *Angular Vein*, receiving branches from the eyelids : from the side of the nose, it passes down over the face, at some distance from the corner of the mouth ; it is here named the *Facial Vein*, and approaches the Facial artery, and then passing over the jaw into the neck, close to the outer side of the artery, usually terminates in the Internal Jugular.

The *Temporal Vein* is seen on the side of the head ; it descends close by the temporal artery, over the zygoma, receiving corresponding branches, and passes beneath the parotid gland. The *Internal Maxillary Vein* exactly resembles the artery of the same name, in the number and distribution of its branches. These two veins unite in front of the ear, the trunk being seated behind the Parotid, but more superficial than the external carotid artery.

It is by the union of the temporal and internal maxillary Veins, and a posterior auricular branch, that the *External Jugular Vein* is chiefly formed, which was seen descending as a sub-cutaneous vein in the neck.

III. NERVES.

1. The PORTIO DURA of the seventh pair, or FACIAL NERVE *, after its course through the aqueduct of Fallopius, comes out by the foramen stylo-mastoideum. The trunk of this nerve is to be sought for by dissecting deep on the anterior margin of the mastoid process, in the hollow behind the ramus of the lower jaw. Make your incision along the anterior border of the sterno-cleido-mastoideus, lift up the edge of the parotid gland, and you will find the nerve, deeply seated at the bottom of this hollow, and immediately passing into the substance of the parotid. Just after its exit from the stylo-mastoid foramen, it sends off the *auricular branch*, behind the ear, and the *nervus stylo-hyoideus* and *N. digastricus*, to the muscles so named. The facial nerve is here close to the posterior auricular artery;—having entered the parotid gland, it lies at first deeply buried in its substance, crossing the line of the external carotid; then descending obliquely forwards, and towards the surface, it divides after a course of little more than half an inch, while yet deeply-seated in the gland, into two branches †. (1.) The *Superior*, or *temporo-facial branch*,

* *Lesser Sympathetic Nerve. Respiratory Nerve of the Face* of Mr. Bell.

† This disposition of the facial nerve is not constant; frequently,

proceeds upwards and forwards in the substance of the parotid, and divides into seven or eight filaments, which, diverging, give the appearance termed *Pes Anserinus* *, and spread themselves on the temple and face: these branches are distinguished into *temporal, malar,* and *buccal.* (2.) The *Inferior* or *Cervico-facial branch* descends obliquely in the substance of the Parotid, to the angle of the jaw, and subdivides into branches, of which some pass over the jaw and masseter, to the side of the face, under lip, and chin: others descend under the platysma myoides in the neck.

The Facial nerve is remarkable for its plexiform arrangement, and the frequent communication of its filaments with each other, and with the branches of the fifth pair and other nerves.

2. The SUPERIOR CERVICAL NERVES send off some ascending branches to the side of the face and head; these are chiefly distributed to the integuments, and communicate with the branches of the portio dura.

As, in the course of this dissection, we meet with many filaments of the second and third branches of the fifth pair of nerves, it will be advisable here to describe these nerves.

3. The SUPERIOR MAXILLARY NERVE, or SECOND BRANCH of the FIFTH PAIR, having left the cranium by the foramen rotundum of the sphenoid bone, emerges behind the tuber maxillare at the lower back part of the orbit, and at the root of the pterygoid process of the sphenoid bone: it here passes transversely through the space, termed the spheno-maxillary fossa, and enters the infra-orbitar canal, through which it is continued, be-

instead of subdividing into two primary branches, it spreads out widely into ascending, transverse, and descending filaments.

* *Plexus Parotideus.*

neath the floor of the orbit, to the face. Its branches, come off before it enters the canal.

(1.) The *Orbitar*, a small branch, which passes into the orbit by the spheno-maxillary fissure, and subdivides into a *temporal* and *malar* branch: the latter, after passing through a canal in the os malæ, becomes cutaneous on the cheek. (2.) *Two short branches*, which unite and form the *spheno-palatine Ganglion*, or *Ganglion of Meckel* :—this small triangular ganglion is buried in the adipose tissue, close to the spheno-palatine foramen, and is not easily discovered *. It gives off many branches, viz. (*a*) the *Spheno-palatine*, or *Nasal* nerves, three to five in number, which pass through the spheno-palatine foramen to the nasal fossæ; of these, one termed *Naso-palatine*, passes to the anterior palatine canal, terminating, with its fellow, in a small *ganglion*, before reaching the mouth. (*b*) *Palatine nerves*, three in number, to the palate and tonsils, of which the largest enters the posterior palatine canal. (*c*) The *Vidian* or *Pterygoid Nerve*, which runs backwards through the pterygoid canal, and piercing the cartilaginous plate between the sphenoid and petrous bones, divides into two filaments, of which one passes into the carotid canal, to unite with the filaments of the superior cervical ganglion; —the other, entering the cranium, runs, beneath the dura mater, in a groove on the upper surface of the petrous bone, to join the portio dura † in the aqueduct of Fallopius. (3.) The *Superior Dental*, or *Alveolar Nerve*,

* This Ganglion does not always exist; when it is absent, the filaments come off from a *pterygo-palatine* branch, or subdivision of the superior maxillary trunk.

† Most anatomists now agree, that no anastomosis really takes place; the Vidian nerve being continued across the tympanum, as the *chorda tympani*, and emerging by the glenoidal fissure.

consists of three or four filaments, which descend along the posterior surface of 'the tuber maxillare, and pass through small canals to the antrum and teeth, supplying also the gums and buccinator. (4.) The *Infra-orbitar Nerve* is the continued trunk of the superior maxillary; in its canal, it gives off the *anterior dental nerves*, and then issues from the infra-orbitar foramen, ramifying widely on the skin and muscles of the cheek, lower palpebra, outside of the nose, and upper lip.

4. The INFERIOR MAXILLARY NERVE, or THIRD BRANCH of the FIFTH PAIR, leaves the cranium by the foramen ovale of the sphenoid bone; the two portions of which the nerve is formed, are distinguishable outside the skull. It is the largest of the divisions of the fifth pair, and usually divides in the zygomatic fossa into two branches *.

From the Superior Branch proceed (1.) The *Deep Temporal Nerves*, usually two in number, to the temporal fossa and muscle. (2.) *Ramus Massetericus*, which passes between the temporal muscle and the neck of the lower jaw, to the inner surface and substance of the masseter. (3.) *Ramus Buccalis* passes downwards and forwards between the two pterygoid muscles, then between the coronoid process and buccinator to the cheek. (4.) The *Pterygoid Nerve*, one or two very delicate filaments to the pterygoideus internus.

The Inferior Branch, which is much larger, soon subdivides into three branches, (5.) The *Inferior Dental Nerve*, a large branch, which first descends between the two pterygoid muscles, and then between the pterygoideus internus and inferior maxilla. It gives off its *Ramus mylo-hyoideus* downwards to the mylo-hyoideus,

* This bifurcation is not constant; sometimes the nerves come off from the trunk.

and adjacent muscles, and submaxillary gland, and then enters the *Dental canal* at the angle of the lower jaw, giving nerves to the teeth, and emerging by the foramen mentale on the chin. (6.) The *Lingual* or *Gustatory Nerve* is destined to the tongue; it first passes downwards, and receives the *chorda tympani* through the glenoidal fissure, then descends between the internal pterygoid muscle and the jaw, and turns forwards to reach the side of the tongue. It was seen in the dissection of the neck, lying close upon the jaw-bone, above the superior fibres of the mylo-hyoideus. (7.) The *Auricular*, or *superficial temporal branch*, passes outwards between the condyle of the lower jaw and auditory meatus, and emerges from beneath the parotid gland, at the root of the zygoma; it divides into two branches, which are distributed, with the temporal arteries, to the side of the head *.

5. In the dissection of the face, we also meet with the FRONTAL or SUPRA-ORBITAR Nerve, which comes from the first branch of the fifth pair; its chief branch emerges through the supra-orbitar foramen: the lesser or *inner frontal* branch passes above the trochlea of the oblique muscle. These nerves are distributed to the forehead.

* To expose the second and third divisions of the fifth pair, saw through the zygoma near its root, divide the os malæ at its junction with the frontal and maxillary bones, and raise the intervening bony arch, the side of the maxilla inferior having been turned outwards. The inferior maxillary nerve is found in the bottom of the zygomatic fossa, and its two largest branches, the dental and lingual, are readily discovered. The superior maxillary nerve is deeply buried in the loose fat of the spheno-maxillary fossa, crossing to the posterior opening of the infra-orbitar canal. To dissect these nerves more fully, (with the first branch of the fifth pair,) the base of the skull, forming the external and upper walls of the orbit, must also be broken up. Steeping in dilute acid softens the bones, while it gives firmness to the medulla of the brain and nerves.

CHAPTER X.

DISSECTION OF THE NOSE, MOUTH, AND THROAT.

SECTION I.

OF THE NOSE.

THE EXTERNAL PART, or ARCH of the NOSE, is formed by the ossa nasi, and four cartilages, two on each side, viz. the LATERAL CARTILAGES, and the CARTILAGES OF THE ALÆ NASI. The former, of triangular shape, are placed on the side of the nose, below the ossa nasi, and between the ascending processes of the superior maxillæ, uniting anteriorly with the cartilage of the septum; while the cartilages of the alæ, of an irregular oval figure, form the aperture and alæ of the nostrils. These cartilages are connected with each other, and with the edges of the ossa nasi and maxillary bones, by ligamentous fibres, in which some detached portions of cartilage are seen. Externally, they are covered by the integuments and muscular fibres; internally, by the lining membrane of the nose.

To display the internal parts of the nose, make a perpendicular section of the face; the saw may be applied on one side of the crista galli, and carried through the base of the skull, and bones of the face, into the cavity of the nose, on one side of the septum narium. Continue the section through the palate into the mouth. The internal part of the nose, *Internal Nares*, or *Nasal Fossæ*, are now exposed,—lined by the pituitary membrane, of an irregular quadrilateral figure, divided into two cavities by the septum, having two openings on the face, and terminating posteriorly by two oval openings, termed the POSTERIOR NARES.

The SEPTUM NARIUM is seen, extending downward from the skull, formed above by the nasal lamella of the ethmoid bone, below and behind by the vomer, and anteriorly by the CARTILAGE OF THE SEPTUM, which is triangular, and, uniting with the front edge of the lateral cartilage on each side, forms the dorsum of the nose.

The external portion of the section, forming the outer boundary of the internal nares, presents an irregular surface, with three projecting portions, named the TURBINATED BONES, of which the upper one is small. The intermediate grooves or smooth passages are termed the MEATUS NARIUM. (1.) The *Superior Meatus*, narrow, with one or two openings of the ethmoidal cells, and sphenoidal sinus: the spheno-palatine foramen also corresponds to its back part, transmitting nerves and vessels to the pituitary membrane. (2.) The *Middle Meatus*, in which we perceive the narrow opening of the great maxillary sinus, and another, which is common to the frontal sinuses and anterior ethmoidal cells. (3.) The *Inferior Meatus*, between the inferior turbinated bone and floor of the nostril, is the largest, in the fore part of which, under the turbinated bone, is found the oblique, slit-like opening of the NASAL DUCT, or CANAL, with its minute valvular plica or fold.

The extent and boundaries of the MAXILLARY SINUS, or ANTRUM, should be examined; for this purpose the superior maxillary bone may be cut through in a line below the orbit. The orifices of the EUSTACHIAN TUBES should also be noticed, on the back part of the pharynx, opposite to the middle meatus of the nose.

The Nasal fossæ are lined by a mucous membrane, which has received the name of the *Pituitary*, or

Schneider's membrane. In the nares, this lining is of considerable thickness, consisting of two closely united laminæ; the outer, adhering to the bones, fibrous, and serving as periosteum ;—the inner, thick, soft, and distinctly villous : but, where it is prolonged into the accessory cavities, or *Sinuses*, it becomes thinner, smooth, and much less vascular. This membrane is continued, posteriorly, with the lining of the fauces; with the conjunctiva of the eye, through the nasal duct; by the Eustachian tube, with the lining of the tympanum, and, at the opening of the nostrils, with the external skin.

SECTION II.

OF THE MOUTH AND THROAT.

THE MOUTH is bounded by the lips anteriorly ; behind by the opening into the throat ; above, by the palate; below, by the tongue ; and laterally by the cheeks. It is lined with a mucous membrane, which is continuous with the mucous membrane of the pharynx, and pituitary lining of the nose, and, along the margin of the lips, with the skin. Numerous small mucous glands are found beneath this lining membrane.

On looking into the mouth, we observe a soft curtain hanging from the palate bones, named the VELUM PENDULUM PALATI, or soft Palate. The apex of the velum forms a small projecting conical body, termed the UVULA, or pap of the throat. From each side of the uvula, two half-arches or columns are sent down, the anterior to the root of the tongue, the posterior to the side of the pharynx : these are formed by a fold of the mucous membrane, and some muscular fibres found beneath it. Between these half-arches on each side, are situated the oval glandular bodies, termed AMYGDALÆ or TONSILS, these consist apparently of a mass of mucous follicles, united by a pulpy tissue, and opening into cellules or orifices, which are large and visible. The common opening behind the anterior arch is named the FAUCES, ISTHMUS FAUCIUM, or top of the Throat, from which there are six passages, two upwards, being one to each nostril, the POSTERIOR NARES; two at the sides, the EUSTACHIAN TUBES, passing on each side to the ear; two downwards, of which the anterior is the passage through the GLOTTIS and LARYNX into the trachea; the posterior, which is the largest, is the PHARYNX, or top of the œsophagus, and leads to the stomach.

The TONGUE is seen filling up the space formed by the arch of the lower jaw; it is attached below by its muscles; behind and laterally, chiefly by the mucous membrane reflected over it. It is divided into *base* and *apex*, with an upper surface or *dorsum*, and an *inferior surface*: beneath the latter is the *frænum*, which is a fold of the lining membrane, and the orifices of the *submaxillary* and *sublingual ducts*, which should be examined. On the upper surface, we observe a superficial *median furrow*, dividing the tongue into two lateral halves, and terminating posteriorly by a small pit, the *foramen cæcum*: also numerous small prominences, termed *Papillæ*, which have been distinguished into three kinds, from their appearance; *Capitatæ*, *Lenticulares*, and *Conicæ*.

The substance of the tongue consists of muscular fibres, interlaced in a confused manner. In the centre of the organ, a denser line of tissue may be observed, placed in the median line, and having its lower part extended between the genio-glossi muscles. This is described by some, as a *Median Cartilage* or *Septum*, giving attachment to muscular fibres by its lateral surfaces.

The PHARNYX is a funnel-like cavity, deficient anteriorly, forming the first portion of the alimentary canal. It lies in front of the spinal column, close upon the muscles covering the cervical vertebræ, being connected with them by loose cellular membrane, and extending downwards behind the larynx. It is bounded above by the base of the skull; anteriorly, by the posterior nares, isthmus of the fauces, base of the tongue, and top of the larynx ; and, below, it terminates in the œsophagus, opposite to the beginning of the trachea, and about the fifth cervical vertebra. The pharynx has a mucous lining, prolonged from the mouth and fauces, external to which is a layer of muscular fibres, investing the sides and back part of the bag, and shortly to be described.

The EPIGLOTTIS is seen projecting at the base of the tongue, into the fore part of the pharynx ; it covers the opening of the larynx.

SECTION III.

OF THE MUSCLES SITUATED ABOUT THE ENTRY OF THE FAUCES, AND ON-THE POSTERIOR PART OF THE PHARYNX.

To display these parts, remove the whole of the lower jaw ; distend the pharynx with horse-hair or tow, and proceed with the dissection : but to expose more fully the constrictor muscles of the pharynx, it is convenient to cut across the trachea and œsophagus above the sternum, so as to admit of the parts being raised from the spine. If the muscles about the back part of the neck are not required for examination, we may divide the spinal column and soft parts close to the occipital bone, and remove the cranium with the pharynx and larynx attached to its base.

§. 1. MUSCLES SITUATED ABOUT THE ENTRY OF THE
FAUCES.

These consist of four pairs, and a single muscle. They are enclosed within the velum and its dependent arches, and are to be exposed by carefully reflecting the mucous membrane.

1. CONSTRICTOR ISTHMI FAUCIUM, or PALATO-GLOSSUS, is a slender muscle, *arising* from the side of the tongue near its root; thence running upwards within the anterior half arch, before the tonsil, it is

Inserted into the middle of the velum pendulum palati, as far as the root of the uvula. It is here connected with its fellow, and with the beginning of the palato-pharyngeus.

Situation: It forms the anterior half-arch of the palate.

Use: To depress the velum towards the root of the tongue, which at the same time it raises, so as to contract the opening into the fauces.

2. The PALATO-PHARYNGEUS—*Arises*, by a broad beginning, from the middle of the velum palati, at the root of the uvula, and from the tendinous expansion of the circumflexus palati. The fibres pass along the posterior arch behind the tonsil, and run backwards to the superior and lateral part of the pharynx, where they are scattered, and intermixed with those of the stylo-pharyngeus.

Inserted into the edge of the upper and back part of the thyroid cartilage, and into the back part of the pharynx.

Situation: It forms the posterior half-arch or column.

Use: To depress the soft palate, and raise the pharynx.

3. The Circumflexus, or Tensor Palati—*Arises*, from the root of the pterygoid process of the sphenoid bone, within the foramen ovale, as far back as the spine of that bone; and from the Eustachian tube near its osseous part. From this origin, it runs down along the inside of the pterygoideus internus, and forms a slender round tendon, which passes over the hook of the internal plate of the pterygoid process, where it has a small ligament and synovial capsule: it then turns horizontally inwards, and spreads into a broad tendinous expansion.

Inserted into the velum pendulum palati, and semi-lunar edge of the os palati, extending as far as the palatine suture.

Situation: This muscle is seen between the pterygoideus internus, and side of the pharynx. Some of its posterior fibres generally join with the constrictor pharyngis superior and palato-pharyngeus.

Use: To stretch the velum, and draw it downwards.

4. The Levator Palati—*Arises*, tendinous and fleshy, from the extremity of the petrous portion of the temporal bone, and from the cartilaginous part of the Eustachian tube.

Inserted into the middle of the velum pendulum palati, as far as the root of the uvula, uniting with its fellow.

Situation: It is a larger muscle than the last, and is situated on the side of the posterior opening of the nostrils.

Use: To draw the velum upwards, so as to shut the posterior openings of the nostrils.

The single muscle is the

Azygos Uvulæ.—It *arises*, fleshy, from the extremity of the suture which unites the ossa palati, and runs down the whole length of the velum, adhering to the tendons

of the circumflexi palati. It is either single, or sometimes consists of two fasciculi.

Inserted into the tip of the uvula. It is surrounded
by mucous glands.

Use : To raise and shorten the uvula.

§. 2. MUSCLES SITUATED ON THE SIDES AND BACK PART
OF THE PHARYNX.

Of these there are three pairs, forming a thin layer,
applied to the mucous membrane, and partially overlap-
ping each other.

1. The CONSTRICTOR PHARYNGIS INFERIOR—*Arises*
from the outside of the ala of the thyroid cartilage, near
the attachment of the thyro-hyoideus muscle, and from
the side of the cricoid cartilage, near the crico-thyroideus.

Inserted into a white line on the back part of the pharynx, where it is united to its fellow, by a kind of raphe.

Situation : This muscle covers the under part of the
middle constrictor; the superior fibres run obliquely upwards, while the inferior fibres have a transverse direction.

2. The CONSTRICTOR PHARYNGIS MEDIUS—*Arises*
from the cornu of the os hyoides, extending as far forwards as the lesser cornu or appendix, and from the ligament connecting the cornu to the thyroid cartilage.
The superior fibres ascend obliquely, the others run
more transversely.

Inserted into the posterior line of the pharynx, uniting
with the opposite muscle, and, by a pointed ascending
process, into the basilar process of the occipital bone before the foramen magnum.

Situation : The lower part of this muscle is covered
by the muscle last described, while the upper fibres over-

lap the constrictor superior. The insertion of the stylo-
pharyngeus is seen on the side of the pharynx, passing
under the fibres of the constrictor medius.

3. CONSTRICTOR PHARYNGIS SUPERIOR—*Arises* from
the lateral part of the base of the tongue, and from the
inferior maxilla near the last dens molaris; from the internal pterygoid plate, in its lower third,—and from the
pterygo-maxillary ligament, which is extended between
these points, and connects this muscle with the posterior border of the buccinator. The lower fibres are nearly
transverse, the upper are arched upwards to reach the
base of the skull, leaving above them a small portion of
the mucous membrane uncovered.

Inserted into the posterior white line of the pharynx,
and, with the last muscle, into the basilar process by a
pointed aponeurotic process.

Situation: The larger part of this muscle is covered
by the constrictor medius.

Use: Each constrictor muscle compresses that part
of the pharynx which it covers; the middle constrictor
also elevates the os hyoides and larynx, carrying them
backwards; the inferior constrictor slightly elevates the
larynx only. These muscles attach the pharynx to the
several parts, from which their fibres take their origin.

SECTION IV.

OF THE LARYNX AND ITS MUSCLES.

THE LARYNX is composed of five Cartilages.

1. The THYROID CARTILAGE, situated immediately below the os hyoides in the middle of the throat, composed
of two lateral portions or *alæ*, which unite in a middle
prominent line, excavated above; and of four *cornua*,
two on each side. It is attached above by a broad

fibrous membrane *, to the base of the os hyoides, and, on each side, to the cornu of that bone, by a *round ligament*, which extends from the upper cornu of the thyroid cartilage, and generally contains a cartilage. Below, the thyroid cartilage is connected by another fibrous membrane † to the upper edge of the cricoid cartilage; and to each side of that cartilage, by its inferior or lesser cornu, which is smooth at its extremity, and articulates with a similar smooth surface of the cricoid cartilage, and has a synovial capsule. Its posterior surface is concave, enclosing the parts constituting the glottis, and giving attachment to the vocal chords.

2. The Cricoid Cartilage is situated immediately below the thyroid, and has its lower edge united by a fibrous membrane to the first ring of the trachea. This cartilage, in shape, represents a ring, which is much deeper behind than before : on each side, it has a smooth surface, to articulate with the inferior cornu of the thyroid cartilage; posteriorly, its upper border has two articulating surfaces for the arytenoid cartilages.

3. The Epiglottis is a triangular cartilage, very elastic, placed behind the root of the tongue, and covering the entrance of the larynx ; its broader leaf-like part unattached ; its narrow process or pedicle fixed to the notch in the upper part of the thyroid cartilage. The epiglottis is also connected to the parts adjacent by the mucous membrane, which, at the root of the tongue, forms three small folds or *fræna*.

4, and 5. The two Arytenoid Cartilages are situated behind the thyroid cartilage, on the upper edge of the back part of the cricoid cartilage, and between the two alæ of the thyroid. These small cartilages are

---

* *Thyro-hyoid membrane.*   † *Crico-thyroid membrane.*

of a triangular or pyramidal form : their inner surfaces are opposed to each other; and the base of each cartilage is concave and smooth, articulating with the smooth surface of the cricoid cartilage. On the apex, is observed a small moveable cartilaginous body, termed *corniculum laryngis* *.

The two arytenoid cartilages form betwixt themselves and the thyroid a longitudinal fissure, extending from before backwards, which is called the GLOTTIS, or RIMA GLOTTIDIS, and leads to the trachea. Raise the epiglottis, and you perceive the GLOTTIS; at the upper part of this slit, observe on each side a fold of the mucous membrane; these are frequently called the *Superior Ligaments of the Glottis;* and lower down, within the larynx, you remark on each side another fold of this same membrane, on removing which a distinct elastic *ligament* is exposed, extending from the projecting fore part of the base of the arytenoid cartilage, to be implanted, close to its fellow, into the posterior angle of the thyroid cartilage, at some distance below its notch; these are called the *Vocal Chords, Proper,* or *Inferior Ligaments of the Glottis,* and the space or cavity on each side, between the superior and inferior ligaments, is named the *Ventricle* or *Sacculus of the Larynx.* The Larynx is lined with a mucous membrane, and a mass of mucous glands is found about the base of the epiglottis, and in the vicinity of the arytenoid cartilages.

The external muscles of the larynx have been described in the neck. The muscles situated about the glottis consist of four pairs of small muscles and a single one. Dissect off the mucous membrane.

1. The CRICO-ARYTÆNOIDEUS POSTICUS — *Arises,*

---

* *Cartilage of Santorini.*

fleshy, from the whole posterior surface of the cricoid cartilage, by the side of the middle prominent line.

Inserted, narrow, into the back part of the base of the arytenoid cartilage of the same side.

Situation : On the back part of the cricoid cartilage, under the mucous membrane.

Use : To draw the arytenoid cartilage outwards and backwards, opening the rima glottidis.

2. The Crico-Arytænoideus Lateralis—*Arises*, fleshy, from the side of the cricoid cartilage, where it is covered by the ala of the thyroid cartilage.

Inserted into the outer side of the base of the arytenoid cartilage.

Situation : It lies more forward than the last-described muscle, and is seen by drawing the thyroid cartilage from the cricoid.

Use : To open the rima glottidis, by pulling the ligaments from each other.

3. The Thyro-Arytænoideus — *Arises* from the middle and inferior part of the posterior surface of the thyroid cartilage, and runs backwards, to be

Inserted into the fore part of the arytenoid cartilage.

Situation : It is situated more forwards than the last muscle, and is covered by the thyroid cartilage.

Use : To pull the arytenoid cartilage forwards, and thus shorten the ligaments of the glottis.

There is a mass of fibres passing from one arytenoid cartilage to the other, some of which are oblique, others transverse : these are by some considered as one muscle, by others they have been divided into a pair of muscles, and a single one.

4. Arytænoideus Obliquus—*Arises* from the base of one arytenoid cartilage ; and, crossing its fellow, is *inserted* into the tip of the other arytenoid cartilage.

The single muscle is more distinct.—ARYTÆNOIDEUS TRANSVERSUS, which *arises* from the whole length of one arytenoid cartilage, and passes across, to be *inserted* into the whole length of the other arytenoid cartilage. *Situation:* Anterior to the arytænoidei obliqui.

The *use* of these three muscles is to shut the rima glottidis, by bringing the arytenoid cartilages together.

On each side of the larynx, there are also a few muscular fibres, under the mucous membrane, which are named as follows, and are rarely distinct.

I. THYRO-EPIGLOTTIDEUS—*Arising,* by a few pale separated fibres, from the thyroid cartilage, and *inserted* into the epiglottis laterally. 2. ARYTÆNO-EPIGLOTTI-DEUS—*Arising,* by a few slender fibres from the lateral and upper part of the arytenoid cartilage, and *inserted* into the epiglottis, along with the former muscle. *Use:* To expand the epiglottis, and to draw it downwards.

CHAPTER XI.

DISSECTION OF THE EYE AND ITS APPENDAGES.

SECTION I.

OF THE APPENDAGES.

THE EYE-LIDS or PALPEBRÆ are united at their angles, which are termed *Canthi.* On removing the orbicularis palpebrarum, we expose the upper and lower *Tarsi,* or cartilages which give shape to the eye-lids: these cartilages are attached to the edges of the orbits by an expanded fibrous membrane, called the *Ligaments of the Tarsi,* and, at the inner canthus, they are fixed to the bifurcated extremity of the tendon of the orbicularis -

palpebrarum : their posterior surface is lined by the tunica conjunctiva ; and the *Glandulæ Meibomii* are lodged in furrows of this surface, under the conjunctiva : the orifices of these glands are made evident, behind the *Cilia* or eye-lashes, on the edge of the tarsi, by pressing out their sebaceous secretion.

The TUNICA CONJUNCTIVA is seen lining the inner surface of the eye-lids, and reflected from them over the anterior part of the globe of the eye, so that the eye-ball with its muscles is situated behind the conjunctiva. This membrane is red and vascular on the lids, but pale and having few red vessels on the ball; it is loosely connected with the sclerotica, but appears inseparable from the lucid cornea. The conjunctiva is of the class of mucous membranes ; it is continued with the skin along the margin of the lids, and, through the lachrymal puncta and sac, with the mucous lining of the nose.

The CARUNCULA LACHRYMALIS is a small reddish granulated body, placed at the internal angle of the palpebræ, and formed of a mass of mucous follicles. It is covered by the conjunctiva, and, close to it, there is a small crescent-shaped fold of this membrane, called the *Valvula Semilunaris.*

Turn the eye-lids a little outwards, and you perceive the PUNCTA LACHRYMALIA, two small orifices, near the internal angle of the palpebræ, situated one in each eye-lid, in the centre of a small fibro-cartilaginous eminence. These are the orifices of two minute canals, or ducts, (the *Lachrymal Ducts,*) which lead into the LACHRY-MAL SAC. This sac is an oblong membranous bag, or reservoir, situated at the inner angle of the eye, in a depression formed by the os unguis and nasal process of the superior maxillary bone, and immediately behind the tendon of the orbicularis, which crosses the sac trans-

.versely; it is lined by a mucous membrane, and its anterior surface is covered by an aponeurotic layer, which
is fixed to the tendon of the orbicularis, and to the bony
margin of the groove lodging the sac *. The Lachrymal
sac receives the tears by the puncta lachrymalia, and from
the sac they are conveyed into the nose by the NASAL
DUCT or CANAL. This duct, prolonged from the lower
end of the sac, descends obliquely backwards, and opens
into the nose, under the os spongiosum inferius; it is
formed of mucous membrane, enclosed in an osseous
canal, and is from six to eight lines in length. A probe,
with its extremity bent, may be introduced from the
nose through the duct into the lachrymal sac.

Divide the conjunctiva lining the back part of the
upper tarsus, or separate the upper eye-lid from the
orbital margin, and remove the cellular tissue in the outer
and upper part of the orbit; you discover the LACHRY-
MAL GLAND. This gland is situated in a depression of
the os frontis near the temple; is of a yellowish colour,
oval, and of the size of an almond; it adheres to the
periosteum of the orbit, and is separated from the outer
convexity of the ball and the rectus externus muscle by
the adipose tissue, with which it is also surrounded posteriorly. It is composed of small lobules, and sends off
several small ducts, which pierce the tunica conjunctiva

---

* MUSCULUS LACHRYMALIS, or *Horner's Muscle,* is found behind
the sac and lachrymal ducts, *arising* from the posterior border of the
os unguis, where it joins the os planum, passing forwards over the
corresponding portion of the lachrymal sac, and bifurcating to be lost
on the two lachrymal canals or ducts. *Use:* To draw the puncta
inwards, adapting the eye-lids to the ball, and perhaps to compress
the lachrymal sac, as suggested by Trasmondi, who has traced two
filaments from the nasal branch of the ophthalmic nerve to this
muscle. This little fleshy mass, discovered by the American Pro-
fessor, is very distinct.

lining the, upper eye-lid; these ducts cannot be seen, unless the part be macerated in coloured water, when they are filled with the liquid.

SECTION II.

OF THE MUSCLES SITUATED WITHIN THE ORBIT.

The Globe, or ball of the eye, is situated about the middle of the orbit, it is connected to the bone by its muscles and by the optic nerve; and all these parts are embedded posteriorly in a soft, fatty substance, which fills up the bottom of the orbit. The eye-ball lies behind the conjunctiva, which must therefore be dissected away; and the upper part or roof of the orbit, which is formed by the os frontis, must also be removed. Saw through the orbitar process near the outer and inner angles, and remove the intermediate portion of bone; taking care to leave the optic foramen, where the muscles have their origin, and the trochlea or pulley of the superior oblique muscle. Dissect away cautiously the loose fat with the scissors.

Seven muscles are contained within the orbit, of which one belongs to the upper eye-lid, and six to the globe of the eye.

The single muscle belonging to the upper eye-lid, is

The Levator Palpebræ Superioris.—It *arises*, by a small tendon, from the upper part of the foramen opticum, and proceeds forwards, forming a broad flat belly,

Inserted, by a broad thin tendon, into the upper eye-lid, adhering to the upper edge of the tarsal cartilage, nearly along its whole length.

Situation : Posteriorly, the upper surface of this muscle is applied to the roof of the orbit; more forwards it is covered by adipose tissue, and anteriorly it is separated, by the ligament of the tarsus, from the fibres of the orbicularis palpebrarum. Its under surface is in contact, behind, with the rectus superior, and, anteriorly, with the tunica conjunctiva.

Use : To open the eye, by drawing the superior eyelid upwards.

The muscles of the globe of the eye are six in number.

Of these there are four straight muscles, or recti, all arising by narrow tendons from the margin of the foramen opticum, where they surround the optic nerve ; all forming strong fleshy bellies, and inserted, by broad thin tendons, at the fore part of the globe of the eye, into the tunica sclerotica, and under the tunica conjunctiva. These straight muscles are also united by cellular tissue with one another, and form a kind of conical sac for the eye-ball, separating it from the bony walls of the orbit, —and the inferior, external, and internal recti, have one common tendinous origin : their tendinous expansions adhere firmly to the sclerotica, and become identified with it.

I. The RECTUS SUPERIOR, or LEVATOR OCULI—*Arises*, by a narrow tendon, from the upper part of the foramen opticum, immediately below the levator palpebræ; it forms a flat belly, which is applied to the upper part of the ball of the eye.

Inserted, by a broad thin tendon, into the sclerotica, about two lines behind the cornea.

Use : To raise the globe of the eye.

2. The RECTUS INFERIOR, or DEPRESSOR OCULI— *Arises*, by the common tendon, from the inferior margin of the foramen opticum, and extends beneath the optic nerve, to be

Inserted into the sclerotica, like the rectus superior.

Use : To move the globe of the eye downwards.

3. The RECTUS INTERNUS, or ADDUCTOR OCULI—
Arises from the inner part of the foramen opticum, by
the common tendon, and is

Inserted into the inner part of the sclerotica.

It is the shortest of the four recti muscles.

Use : To draw the eye towards the nose.

4. The RECTUS EXTERNUS, or ABDUCTOR OCULI—
Arises from the outer part of the foramen opticum, by
the common tendon, adhering intimately to the rectus
inferior,—and, by a *second* slip of tendon, from the body
of the os sphenoides, uniting with the tendon of the
superior rectus.

Inserted into the outer part of the sclerotica.

It is the longest of the recti. The bifurcated origin
of this muscle gives a passage to the third and sixth
nerves, and nasal branch of the ophthalmic.

Use : To move the globe outwards. When the recti
act together, they retract the eye into the orbit.

The two next are oblique muscles.

5. The OBLIQUUS SUPERIOR, or TROCHLEARIS—
Arises, by a small tendon, from the margin of the foramen opticum, at its upper and inner part. Its long
slender belly runs along the inner side of the orbit to the
internal angular process of the os frontis, where a cartilaginous ring or pulley is fixed. The muscle here forms
a rounded tendon, which passes through the pulley, runs
obliquely downwards and outwards, enclosed in a membranous sheath ; and then passes between the eye-ball
and rectus superior, becoming broader and thinner.

Inserted into the upper and back part of the sclerotica, about mid-way between the insertion of the rectus
superior, and the entrance of the optic nerve.

Use : To roll the globe of the eye, and turn the pupil
downwards and outwards.

6. The Obliquus Inferior—*Arises,* narrow, and principally tendinous, from the outer edge of the orbitar process of the superior maxilla, a few lines external to the lachrymal sac. It runs obliquely outwards and backwards, under the inferior rectus, and then turning upwards between the globe and external rectus, is

Inserted, by a broad thin tendon, into the outer part of the sclerotica, nearer to the optic nerve than the insertion of the superior oblique.

Use : To carry the eye inwards and forwards, and to turn the pupil upwards. The two obliqui are also antagonists to the recti, opposing the retraction of the eye into the orbit.

———

SECTION III.

OF THE VESSELS AND NERVES MET WITH IN THE ORBIT OF THE EYE.

I. ARTERIES.

THE vessels in the orbit cannot be readily traced, unless they are injected.

The OPHTHALMIC ARTERY is a branch of the internal carotid. It enters the orbit by the foramen opticum ; it is at first placed below and on the outer side of the nerve ; it then ascends, and crosses obliquely over it, being covered by the rectus superior, and continues its course along the inner side of the orbit to its internal angle, where it terminates by dividing into two branches.

The branches of the ophthalmic artery are numerous :

1. The *Lachrymal Artery,* to the lachrymal gland and parts adjacent.

2. The *Central Artery* of the *Retina,* which penetrates the coats of the optic nerve, and is continued through

its centre to the eye, where it is distributed to the retina.

3. The *Supra-Orbitar*, or *Superciliary Artery*, which passes forwards along the roof of the orbit, above the levator palpebræ and rectus superior, to the superciliary notch, and is continued through the notch to the muscles and integuments of the forehead.

4. The *Ciliary Arteries*, divided into *(a) A. ciliares breves*, twenty or more in number, penetrating the sclerotic near the optic nerve : *(b) A. ciliares longæ*, two in number, penetrating the sclerotic, on each side, a line or two more forwards. *(c) A. ciliares anteriores*, four or five in number, commonly derived from the muscular branches, and piercing the sclerotic near the cornea. These arteries are distributed chiefly to the choroid coat and iris.

5. The *Superior* and *Inferior Muscular Arteries*, to the muscles, and periosteum of the orbit : the superior is sometimes wanting.

6. The *Anterior* and *Posterior Ethmoidal Arteries*, which enter the cranium through the internal orbitar foramina, and then pass through the cribriform lamella into the nose.

7. The *Superior* and *Inferior Palpebral Arteries*, to the orbicularis muscle, and neighbouring parts.

8. The two branches in which the ophthalmic artery terminates, are *(a)* the *Nasal Artery*, which leaves the orbit above the tendon of the orbicularis palpebrarum, and passes on the side of the root of the nose, giving off twigs, and inosculating with the extreme branches of the facial artery. *(b)* The *Frontal Artery*, which leaves the orbit rather higher up, and ascends on the forehead, between the bone and orbicularis.

The INFRA-ORBITAR ARTERY is found in the lower

part of the orbit; it is the continued trunk of the internal maxillary, and is seen passing along its canal, in the upper or orbitar plate of the superior maxilla, to emerge on the face by the infra-orbitar foramen.

II. VEINS.

These correspond with the arteries; they discharge their blood partly into the branches of the facial vein near the forehead and temples, and partly into the oph-thalmic vein, which passes through the foramen lace-rum into the cavernous sinus.

III. NERVES.

1. The OPTIC NERVE is seen coming through the fo-ramen opticum, and proceeding forwards, surrounded by the four recti muscles, to enter the back part of the globe of the eye, and form the retina. It is invested by a dense sheath, prolonged from the dura mater.

2. The Nerve of the Third Pair, MOTOR OCULI, comes into the orbit through the superior orbitary fissure, or foramen lacerum *, passing between the two heads of the external rectus muscle; it is divided into two branches, which pierce separately the dura mater closing the fis-sure.

(1.) The superior and smaller branch runs upwards, passing above the optic nerve, and subdivides into two nerves, of which one supplies the rectus superior, and the other the levator palpebræ superioris.

(2.) The inferior and larger branch passes below and to the outside of the optic nerve, and subdivides into three branches, which go to the rectus inferior, rectus internus, and obliquus inferior; the branch, which sup-

* *Sphenoidal Fissure.*

plies the obliquus inferior, is long, and gives off a filament to the lenticular ganglion.

3. The Nerve of the Fourth Pair, N. PATHETICUS, or Trochlearis, enters the orbit by the superior orbitary fissure; this slender nerve then inclines upwards and inwards, close to the periosteum, and runs to the obliquus superior.

4. The first branch of the Nerve of the Fifth Pair, named OPHTHALMIC, also enters the orbit by the superior orbitary fissure; it is divided into three branches, which pierce the dura mater separately.

(1.) The *Lachrymal Branch* supplies the lachrymal gland, and the parts at the external angle of the palpebræ.

(2.) The *Frontal* or *Supra-Orbitar Nerve* is the largest branch; it is at first connected with the fourth pair, passing upwards above the levator palpebræ, close along the upper wall of the orbit;—then proceeding through the superciliary notch or foramen, it is distributed to the forehead:—it gives off a branch, which passes out of the cavity, on the inner side, near the pulley of the obliquus superior, and ramifies to the upper eye-lid and forehead; this is called the *Inner Frontal* or *Supra-trochlear Nerve.*

(3.) The *Nasal Nerve* or inner branch, enters the orbit between the two origins of the external rectus muscle; it is then directed inwards, passing between the optic nerve and rectus superior, and reaches the internal side of the orbit: it gives off a filament to the lenticular ganglion, and subdivides into two branches, of which *(a,)* the *internal nasal branch* passes through the foramen orbitarium internum anterius into the skull, and then through the cribriform lamella into the nose. *(b,)* The *external nasal* or *infra-trochlear branch* follows the original direction of the nerve, and passes out of the orbit below the pulley of the obliquus superior ; it is distributed to the eye-lids, lachrymal sac, and integuments of the nose and forehead.

The LENTICULAR or OPHTHALMIC GANGLION, is a small ganglion situated within the orbit, formed by the filament of the nasal branch of the ophthalmic nerve, and the filament supplied by the third pair. It is found, in the orbit on the outside of the optic nerve, between the nerve and external rectus, surrounded by fat and cellular tissue ; it varies in size, and its colour is white or reddish ; it sends off delicate nerves, which run along the sides of the optic nerve, and pierce the coats of the eye, to supply the iris.

5. The nerve of the Sixth Pair, ABDUCTOR OCULI, also enters the orbit through the orbitary fissure, between the two heads of the external rectus, and beneath the other ocular nerves ; it is distributed to the external rectus, piercing its inner surface.

6. The INFRA-ORBITAR NERVE, or terminating branch of the Superior Maxillary, is seen crossing the floor of the orbit, lying at first half-exposed in its groove, which then becomes a canal, and transmits the nerve, with the artery of the same name, to emerge on the face by the infra-orbitar hole.

These delicate nerves are surrounded by the adipose substance found in the orbit, and require to be dissected with the utmost care.

SECTION IV.

DISSECTION OF THE GLOBE OF THE EYE.

THE Eye-ball must be removed from the orbit, and cleared of its muscles and fat. First observe the dense exterior coat of the eye, composed of two portions of very dissimilar appearance. (1.) The SCLEROTICA, occupying the posterior four-fifths of the ball, fibrous, opaque, and of a pearl white colour. (2.) The CORNEA, adhering firmly to the sclerotica, but separable by maceration, constituting the anterior fifth part, convex, transparent, and divisible into lamellæ. Remark that the sclerotica is pierced behind by the optic nerve, and that the fibrous sheath derived from the dura mater, which invests the nerve, is continued into the sclerotica.

Remove the cornea by a circular incision,—the *aqueous humour* will escape from its chamber or cavity behind the cornea. Placed transversely across this cavity, you observe a circular flattened partition, with a central aperture: this is the IRIS, adhering by its circumference to the ciliary ligament, and forming by its inner border the PUPIL. The aqueous humour having escaped, the Iris now subsides on the transparent body behind it, the CRYSTALLINE LENS, which is lodged in the anterior part of the vitreous humour: pass the point of your scalpel under the iris, and examine its form and situation: scratch the capsule of the lens, and press on the ball,—the lens will escape from the pupil, and, on further pressure, will be followed by the VITREOUS HUMOUR. Slitting up the sclerotica, you see the RETINA and CHOROID COAT: the former is a pulpy pellicle, easily detached from the inner surface of the choroid, which appears lined with a black crust, or *pigment:*—the choroid is placed immediately within the sclerotica.

Take another eye : puncture the sclerotica a few lines behind the cornea, and introduce cautiously one blade of the scissors : divide a portion of sclerotica, and raise it from the dark-coloured delicate membrane beneath. This is again the CHOROID, exposed now on its outer surface, and its colour is owing to the *nigrum pigmentum* lining it internally, a thinner layer of which pigment is also found on this outer surface. The dissection should now be continued on the eye placed under water in a shallow basin * ; extend the incision of the sclerotica circularly, and detach it from the choroid; on the outer surface of the latter, some longitudinal filaments, the *Ciliary Nerves*, are seen running forwards to the iris ; and many small veins, the *vasa vorticosa*, are observed, if distended, disposed in diverging arches or *vortices*.

Continuing to reflect the sclerotic towards its junction with the cornea, you find its inner surface adherent to a firm *greyish ring*, a line or two in breadth, extending round the eye ; this ring appears to form the anterior boundary of the choroid, and is closely united with it : it is termed the CILIARY LIGAMENT, and its inner circumference will be found to give attachment to the iris.

Raise a portion of the choroid with the forceps, and detach it from the retina beneath, which is easily done, as there is no adhesion, the two membranes being simply contiguous : the inner surface of the choroid, or *Tunica Ruyschiana*, is again seen, lined with its black pigment. The RETINA is now exposed, extending as a thin greyish

* This is for the purpose of giving support to the delicate internal membranes and humours during the examination. It is also necessary to fix the eye, which is effected by attaching it to a piece of cork, wax, &c. by means of threads carried through the sclerotica and pins fixed to the cork ;—or a pin may be pushed across the cornea for the attachment of ligatures.

medullary expansion over the vitreous humour : it consists of two layers, (1) a medullary *external* layer, which
is pulpy, and may be scraped off, or separated by maceration, from (2) the *internal* layer, consisting chiefly of
vessels; a *third cellular membrane* has been recently
discovered by Dr. Jacob, between the retina and choroid.
Enlarge the section of the choroid and sclerotica, remove the eye from the water, and allow the humours to
drop out. The Retina falls inwards from the choroid;
it may be made more opaque by a weak acid : it commences, behind, around the bulbous extremity of the
optic nerve; and is evident as far forwards as the ciliary
processes, where it appears to terminate, but it may be
traced between these processes as far nearly as the lens.
On the inner surface of the retina, about two lines on
the outer side of the optic nerve, is the *yellow spot of
Sœmmering*, with its minute central foramen : this appearance is situated directly in the axis of the eye, and
is most distinctly seen by making a transverse section
of a fresh eye.

Now inverting the eye partially, you observe that the
anterior margin of the choroid is thrown into folds, which
lie loose, and form a black radiated ring behind the
iris. These are the CILIARY PROCESSES ;—their situation
is between the iris and vitreous humour : these folds are
of a triangular shape, they extend inwards from the ciliary ligament, and form by their union a circular ring, the
Corpus Ciliare, around the crystalline lens, being lodged
in corresponding depressions on the fore part of the vitreous humour. Anterior to these processes, you observe
the IRIS, adhering by its outer circle to the ciliary ligament ; its posterior surface is also covered with the black
pigment continued from the inner surface of the choroid,
and is named *Uvea*. In the posterior half of the sec-

tion of the eye, you may remark the entrance of the optic nerve, which, in passing through the sclerotic, is much contracted in size ; on squeezing the nerve, the medulla protrudes through several small foramina, which form the *Lamina Cribrosa* of the sclerotic : the *porus Opticus* is a small foramen in the nerve, marking the place of the arteria centralis retinæ.

You may now examine the mass of humours, which escaped from the eye. It consists of a soft gelatinous mass, the VITREOUS HUMOUR, having on its anterior part a middle depression, in which a transparent lenticular body, the CRYSTALLINE LENS is lodged. The vitreous humour is enclosed in a transparent capsule, the *Hyaloid Membrane*, which sends processes inwards, forming cells to contain the humour : these cells communicate, and on puncturing one, the whole humour gradually escapes : on the fore part of the hyaloid membrane, black lines are observable, marking where the ciliary processes were lodged.

At the edge of the lens, the hyaloid membrane splits into two layers, one of which is continued into the front part of the capsule of the crystalline, the other passes behind it, leaving a triangular space or interval, termed the *Canal of Petit*. Puncture at the angle between the lens and vitreous humour, and the canal, when inflated, will have a sacculated appearance.

The Crystalline Lens is of lenticular shape, with two convex surfaces, of which the posterior is more prominent : its external plates are soft and gelatinous, with a firmer central nucleus : it has a proper capsule, containing occasionally a minute portion of fluid, the *Liquor Morgagni* :—the Lens varies in transparency and colour according to age.

Various sections of the eye should be made, to obtain

1

a complete knowledge of its structure. The Iris is well displayed by making a circular section of the cornea, and another section behind the iris, and plunging the intermediate portion in water, when the iris is seen floating, with the ciliary processes. The fresh eyes of animals are also advantageously used for this purpose.

CHAPTER XII.

DISSECTION OF THE MUSCLES SITUATED ON THE POSTERIOR PART OF THE TRUNK AND NECK.

From the number and intimate connexion of these muscles, their description is necessarily complicated, and their dissection difficult. The muscles of the external layers are large and distinct, but, on removing these, we shall not find the subjacent muscles, as in the limbs, loosely connected by cellular membrane, and separated with facility, but closely united, and, in many places, having their fibres so intermixed, as to render their divisions indistinct and uncertain. The smaller muscles, indeed, cannot be separated without dividing some of these fibres, and they are also subject to considerable variations.

In this dissection we meet with twenty-two distinct pairs of muscles, besides a number of small muscles situated between the processes of contiguous vertebræ.

The body being turned on the face, a large block is to be placed under the chest, and the arms and head allowed to hang;—then make an incision along the whole length of the spine, from the tuber of the occiput to the coccyx; cross it by another from the upper dorsal vertebræ to the spine of the scapula and acromion; and, beginning from this transverse cut, remove the integuments, upwards and downwards, from the posterior part of the neck and back, following as much as possible the course of the muscular fibres.

This will expose a broad triangular muscle, extending from the occiput, over the posterior part of the neck and shoulder, to the lower part of the back.

1. The TRAPEZIUS—It *arises*, by a thick round tendon, from the lower part of the protuberance in the middle of the os occipitis, and, by a thin tendinous expansion, from the superior transverse ridge of that bone; —from the five superior cervical spinous processes, by the ligamentum nuchæ;—tendinous, from the two inferior cervical spinous processes, and from the spinous processes and inter-spinous ligaments of all the vertebræ of the back. The fleshy fibres coming from the neck descend obliquely, those from the back ascend, while the middle fibres pass transversely, thus converging generally towards the shoulder.

Inserted, fleshy, into the posterior third part of the clavicle; tendinous and fleshy, into the posterior border of the acromion, and into the upper edge of all the spine of the scapula. The fibres slide over a triangular flattened surface at the extremity or base of the spine of that bone, and an indistinct bursa is interposed.

Situation: This muscle is quite superficial, lying immediately under the skin; and the cellular tissue interposed is condensed, and contains but little fat. It conceals all the muscles situated in the posterior part of the neck, and upper part of the back. It adheres to its fellow the whole length of its origin, producing the quadrilateral figure, from which it derives its name. Its anterior fibres lie posterior to those of the sterno-mastoideus, but are not in contact with them, a considerable quantity of adipose substance being interposed.

2

Use : To move the scapula in different directions, carrying backwards the shoulder and clavicle: Also, to draw back the head, and contribute to its rotatory motions.

The *Ligamentum Nuchæ,* or *Superficial Cervical Ligament,* is a narrow ligamentous band, which springs from the middle spine of the occipital bone, and runs down superficially, to be fixed to the sixth or seventh cervical spine, adhering to the intermediate spinous processes, and giving origin, on each side, to the fibres of the trapezius and other muscles.

Under the lower angle of the trapezius, you observe another broad muscle, covering the lower part of the back and whole lumbar region, and extending to the arm.

2. The LATISSIMUS DORSI—*Arises,* by a broad thin tendon, from all the spinous processes of the os sacrum, and of the lumbar vertebræ; from the spinous processes of the six or seven inferior dorsal vertebræ: from the posterior part of the crest of the ilium: also from the extremities of the four inferior false ribs, by four distinct fleshy digitations, which intermix with those of the obliquus externus abdominis. The inferior fleshy fibres ascend obliquely; the superior run transversely: they pass over the inferior angle of the scapula, (from which the muscle often receives a thin fasciculus of fibres,) to reach the axilla, where they are collected and twisted.

Inserted, by a strong flat tendon, into the inner or posterior edge of the bicipital groove of the humerus, the lower border of the tendon receiving the fleshy fibres of the upper part of the muscle, and its upper border those from below.

Situation : This muscle is also of a triangular figure. Where it arises from the dorsal vertebræ, it is concealed by the origin of the trapezius; the remainder of it is placed immediately under the skin, and covers the deeper seated muscles of the loins and back. It is situated superior to the gluteus maximus, and posterior to the obliquus externus abdominis. As it approaches the axilla, it becomes much connected with the teres major, first passing over the lower part of that muscle, then turning under its inferior border, and becoming anterior; and the tendons of the two muscles will be found, in the dissection of the arm, closely united to be fixed to the humerus. It forms the posterior fold of the arm-pit. The broad thin tendon of this muscle, where it comes from the spines of the sacrum and lumbar vertebræ, is closely connected with the subjacent tendon of the serratus posticus inferior, and assists in forming the fascia lumborum.

Use: To draw the arm backwards and downwards, and to rotate the os humeri inwards and backwards. When the arm is fixed, it may raise the four lower ribs; —or the whole trunk, as in climbing.

Remove the trapezius and latissimus dorsi, or reflect them from the spine towards their insertions: two muscles will be seen passing from the neck to the scapula.

3. The RHOMBOIDEUS. This muscle is divided into two portions, by a cellular line.

(1.) Rhomboideus major, (the inferior portion,) *arises,* tendinous, from the spinous processes of the four or five superior dorsal vertebræ, descends obliquely, and is

Inserted into all the base of the scapula below its spine, extending as far as its inferior angle.

(2.) Rhomboideus Minor, (the superior portion,) is much narrower, *arising,* tendinous, from the spinous processes of the three inferior vertebræ of the neck, and from the ligamentum nuchæ.

Inserted into the base of the scapula, opposite to the triangular plain surface at the root of the spine.

Situation: This thin, quadrilateral, muscle lies beneath the trapezius and latissimus dorsi; a small part of the rhomboideus major, however, appearing between these muscles, and the base of the scapula, immediately under the skin :—the middle fibres of the muscle have a peculiar disposition, being inserted into a tendinous arch, which is fixed to the scapula only by its two extremities.

Use: To draw the scapula obliquely upwards and backwards, lowering the shoulder.

4. The LEVATOR SCAPULÆ, or *Angularis—Arises* from the posterior tubercles of the transverse processes of the five first cervical vertebræ, by five distinct tendinous and fleshy slips; frequently only from four or three vertebræ :—these slips unite, and form a considerable muscle, which passes outwards and downwards.

Inserted, tendinous and fleshy, into the base of the scapula, above the root of the spine, and under the superior angle, (not into the angle itself, as it is usually described.)

Situation: This muscle is situated in the lateral and back part of the neck, concealed by the trapezius and sterno-mastoideus: but a small part of its belly may be seen in the space between the edges of these muscles. The origin of the levator scapulæ is partly concealed by the splenius capitis: and the digitations, where they arise from the transverse processes, lie betwixt similar attachments of the scaleni muscles before, and of the splenius colli behind. At its insertion, it is much intermixed with the serratus magnus.

Use: To draw the scapula upwards, and a little forwards; and to rotate it.

Now detaching the rhomboideus from its origin in the spine, or dividing it perpendicularly, you will see another muscle passing from the whole of the base of the scapula,

5. The SERRATUS MAGNUS—*Arises*, by nine fleshy digitations, from the eight or nine superior ribs. These digitations are seen on the anterior part of the thorax; they pass obliquely backwards, and form a strong fleshy muscle.

Inserted, principally fleshy, into the whole of the base of the scapula.

Situation : This broad, flat, muscle lies between the scapula and the ribs, covering the side of the chest, and forming the inner wall of the axillary cavity, so that, to see its course, the articulation of the clavicle with the sternum should be divided, and the scapula lifted from the trunk. It is concealed by the latissimus dorsi, by the two pectoral muscles, and the scapula. The only part of it which can be seen before the removal of those muscles, projects betwixt and below them on the side of the trunk. The lower digitations, which pass more anteriorly than the edge of the latissimus dorsi, are intermixed with the superior digitations of the obliquus externus. The superior digitations arise behind the pectoralis minor. The insertion of the muscle is between the subscapularis, which arises from the internal surface of the scapula, and the insertions of the rhomboideus and levator scapulæ.

Use : To move the scapula forwards, and, when the scapula is forcibly raised, to draw the ribs upwards *.

The removal of the rhomboideus also exposes,

6. The SERRATUS SUPERIOR POSTICUS. This muscle *arises*, by a thin broad tendon, from the spinous processes of the two or three inferior cervical vertebræ, and of the two superior dorsal.

Inserted, by distinct fleshy slips, into the upper border

* The upper extremity may now be removed from the trunk.

and external surface of the second, third, fourth, and sometimes the fifth ribs, a little beyond their angles.

Situation : This muscle is concealed by the rhomboideus and scapula, except a few of its superior fibres, which appear above the upper edge of the rhomboideus minor. It covers part of the origin of the splenius.

Use : To elevate the ribs, and dilate the thorax.

Reflect it from the spine, and observe a strong flat muscle, placed by the sides of the vertebræ, and extending to the occiput.

7. The SPLENIUS :—it is divided into two portions, connected at their origin, but inserted separately.

(1.) The Splenius Capitis, or inner portion—*Arises,* tendinous, from the spinous processes of the two superior dorsal, and five inferior cervical vertebræ, adhering to the ligamentum nuchæ. It forms a flat broad muscle, which ascends obliquely, and is *inserted,* tendinous, into the posterior part of the mastoid process, and into a small part of the os occipitis, below its superior transverse ridge, and immediately beneath the insertion of the sterno-mastoideus.

Situation : This muscle is covered by the trapezius, and by the insertion of the sterno-mastoideus, and a small part of it is seen on the side of the neck betwixt those two muscles. The lower part of its origin is covered by the serratus superior posticus. The two splenii capitis recede from each other as they ascend, and in the triangular space between them is seen a part of the complexus muscle of each side.

(2.) The Splenius Colli, or outer portion—*Arises,* tendinous, from the spinous processes of the third, fourth, fifth, and sometimes the sixth dorsal vertebræ. It forms a small fleshy belly, which ascends by the side of the vertebræ, and is *inserted* into the transverse processes of the three or four superior cervical vertebræ, by separate tendons, which lie behind similar tendons of the levator scapulæ.

Situation: This muscle runs along the outer edge of the splenius capitis, and below is closely united with it: it is concealed by the serratus superior posticus.

Use: To bring the head and upper vertebræ of the neck obliquely backwards, rotating the head and turning it laterally. When the muscles of both sides act, they pull the head directly backwards.

Reflect both portions of the splenii from their origins, and leave them attached by their insertions: and now attend to the lower part of the back, where the latissimus dorsi has been reflected from the spine to its attachment to the ribs. Observe,

8. The SERRATUS POSTICUS INFERIOR—*Arising*, by a broad thin tendon, from the spinous processes of the two or three inferior dorsal vertebræ; and from the three superior lumbar spines, by the fascia lumborum.

Inserted, by distinct fleshy slips, into the lower edges of the four inferior ribs, at a little distance from their cartilages.

Situation: This is a thin muscle, of considerable breadth, situated at the lower part of the back, under the middle of the latissimus dorsi. An extremely thin and transparent APONEUROSIS * may be observed passing from the superior edge of the serratus posticus inferior, to the lower border of the serratus posticus superior: this aponeurosis is also attached to the spinous processes, and to the angles of the ribs, and is therefore of a quadrilateral form.

The tendon of the serratus posticus inferior lies under

---

* *Vertebral Aponeurosis:* it serves to bind down the subjacent deep muscles.

that of the latissimus dorsi, but, although firmly adhering to it, is distinct, and may be separated by cautious dis- section. Its insertion into the ribs is situated imme- diately behind the attachments connecting the latissimus dorsi to the ribs, which attachments must therefore lie between the obliquus externus abdominis, and serratus posticus inferior.

Use : To pull the ribs downwards and backwards.

The FASCIA LUMBORUM is now seen. It is a tendinous fascia, arising from the spinous processes of the lumbar vertebræ, os sacrum, and back part of the crista ilii, and giving origin to the lower part of the serratus posticus inferior, and to the posterior fibres of the obliquus in- ternus, and transversalis abdominis. It is also connected with the tendon of the latissimus dorsi.

On detaching from the spine this fascia, and the ser- ratus posticus inferior, we expose a thick muscular mass, extending from the sacrum upwards on the sides of the lumbar and dorsal vertebræ, and filling up all the space betwixt the spinous processes of the vertebræ, and the angles of the ribs *. This mass consists of three muscles:

(1.) Sacro-lumbalis on the outside,
- (2.) Longissimus dorsi in the middle,
(3.) Spinalis Dorsi close to the spinous processes.

These three muscles are closely connected together †; so that, to effect their separation, it is necessary to divide some of the fibres.

9 and 10. The SACRO-LUMBALIS and LONGISSIMUS DORSI—*Arise,* by one common origin, which is exter-

---

* The proper tendon of the transversalis abdominis is seen passing beneath the lower part of this muscular mass, to be attached to the transverse processes of the lumbar vertebræ, splitting into two laminæ to enclose the quadratus lumborum.

† They form the *Sacro-spinalis* of M. Chaussier.

nally a broad sheet of tendon, but fleshy internally, from the spinous processes and posterior surface of the os sacrum; from the posterior and internal part of the spine of the os ilium, extending nearly as far forwards as the highest part of that bone when the body is erect; and from the spinous processes, and roots of the transverse processes of all the lumbar vertebræ.

The thick fleshy belly, formed by this extensive origin, ascends, and, opposite to the last rib, divides into the two muscles.

The sacro-lumbalis is *inserted* into all the ribs near their angles, by long and thin tendons. . The tendons which pass to the superior ribs are longer, ascend nearly straight, and are situated nearer to the spine, than those tendons which pass to the lower ribs. On separating the inner edge of this muscle, (*i. e.* the edge next to the spine), from the latissimus dorsi, and turning the belly towards the ribs, we see six or eight small tendinous and fleshy bundles, which pass from the inner side of this muscle, to be inserted into the upper edge of the six or eight inferior ribs. These are called the Musculi Accessorii ad Sacro-Lumbalem.

Use : To pull the ribs downwards, to assist in erecting the trunk of the body, and in turning it to one side.

The longissimus dorsi is *inserted* into all the ribs except the two inferior, betwixt their tubercles and angles, by slips which are tendinous and fleshy, and into the transverse processes of all the dorsal vertebræ, by small double tendons. The insertions into the ribs proceed from the outer side of the muscle, while the attachments to the transverse processes are seen on separating the longissimus dorsi from the spinalis dorsi.

Use : To extend the vertebræ, and keep the body erect.

P

The broad sheet of tendon, which covers these two muscles, and from the internal surface of which the fleshy fibres arise, is strong, thick, and of a pearly white : it is penetrated by vessels and nerves, and extends higher on the longissimus dorsi than on the sacro-lumbalis.

11. The Spinalis Dorsi is much smaller than the two last-described muscles ; below, it cannot be separated from the longissimus dorsi, without dividing some fibres ; it lies betwixt that muscle and the spine. It

Arises, tendinous, from the spinous processes of the two superior lumbar vertebræ, and of the two or three inferior dorsal.

Inserted into the spinous processes of the nine upper vertebræ of the back, except the first, by as many distinct tendons.

Use : To extend the spine.

The three last-described muscles are covered below by the serratus posticus inferior and latissimus dorsi : above, by the rhomboideus, serratus superior posticus, and trapezius.

Connected with the upper part of the longissimus dorsi and sacro-lumbalis, we find two narrow muscles ascending to the cervical vertebræ.

12. The Cervicalis Descendens—*Arises* from the upper edge of the four or five superior ribs, near their tubercles, by as many distinct tendons, which lie on the inside of the tendinous insertions of the sacro-lumbalis. It forms a small belly, which ascends upwards, and is

Inserted, by three distinct tendons, into the transverse processes of the fourth, fifth, and sixth cervical vertebræ.

Situation : This muscle (better named, *Ascendens,)* is small ; it is frequently described as an appendage to the sacro-lumbalis. It arises between the sacro-lumbalis and longissimus dorsi, and is inserted into the transverse processes between the splenius colli and levator scapulæ; of course it is concealed by the rhomboideus, &c. It often receives a fleshy slip from the upper part of the longissimus dorsi.

Use : To extend the neck obliquely backwards.

13. The TRANSVERSALIS COLLI aut CERVICIS—*Arises* from the transverse processes of the four or five superior dorsal vertebræ, by as many tendinous and fleshy slips, and proceeds upwards, to be

Inserted, tendinous, into the transverse processes of the five or six inferior cervical vertebræ.

Situation : The origin of this muscle lies on the inside of the longissimus dorsi, and is sometimes considered as an appendage to it. The insertion is situated between the cervicalis descendens and trachelo-mastoideus.

Use : To extend the neck obliquely backwards.

14. The TRACHELO-MASTOIDEUS lies nearer to the spine than the last-described muscle. It *arises* from the transverse processes of the three or four uppermost vertebræ of the back, and of the four or five inferior of the neck, by as many thin tendons, which unite and form a fleshy belly.

Inserted, tendinous, into the posterior surface of the mastoid process.

Situation : This muscle runs from the side of the vertebræ to the mastoid process, close to the outer edge of the complexus, and it lies on the inside of the transversalis colli; its insertion is concealed by the splenius capitis ;—it is covered also by the levator scapulæ.

Use :—To keep the head and neck erect, and to draw the head backwards, and to one side.

15. The COMPLEXUS—*Arises,* by tendinous and fleshy fibres, from the transverse and also from the articular processes of the five or six superior dorsal, and the four

or five inferior cervical vertebræ. It forms a thick, tendinous and fleshy belly, which ascends obliquely.

Inserted, tendinous and fleshy, into the hollow betwixt the two transverse ridges of the os occipitis, extending from the middle protuberance of that bone, nearly as
far as the mastoid process.

Situation : This is a large muscle. Its origin from the
cervical vertebræ is nearer to the spine than the trachelo-
mastoideus, and it arises in the back nearer to the spine
than the transversalis colli ; it is covered by the splenius ;
but a large portion of it is seen between the splenius and
spine, immediately on removing the trapezius. That
portion of the muscle, which is nearest the spine, has a
rounded middle tendon, to each extremity of which the
fleshy fibres are attached, and has been named Biventer
cervicis.

Use : To draw the head backwards, and to one side.

On removing the complexus from the occiput, we find,
close to the spine, a mass of muscle filling up the space
between the spinous and transverse processes, from the
second cervical vertebra to the middle of the back :—
this is

16. The SEMI-SPINALIS COLLI.—It *arises*, by distinct
tendons, from the transverse processes of the five or six
superior dorsal vertebræ, ascends obliquely close to the
spine, and is

Inserted into the spinous processes of all the vertebræ
of the neck, except the first and the last.

Situation : This muscle is situated close to the vertebræ at the posterior part of the neck and back. It
arises on the outside of the semi-spinalis dorsi ; its
greater part is concealed by the complexus and longissimus dorsi ; and the part which projects between these
muscles, is concealed by the serratus superior posticus.

Use : To extend the neck obliquely backwards.

17. SEMI-SPINALIS DORSI—*Arises* from the transverse processes of the seventh, eighth, and ninth vertebræ of the back, by distinct tendons which soon grow fleshy.

Inserted, by distinct slips, into the spinous processes of the five superior dorsal vertebræ, and of the two lower cervical.

Situation : This muscle lies nearer the spine than the lower part of the semi-spinalis colli; its inferior origins lie on the outside of the insertion of the spinalis dorsi.

Use : To extend the spine obliquely backwards.

The removal of the complexus brings also into view four small muscles, situated at the superior part of the neck, immediately below the occiput, and extending to the two first vertebræ.

18. The RECTUS CAPITIS POSTICUS MAJOR—*Arises,* fleshy, from the side of the spinous process of the dentata, or second cervical vertebra. It ascends obliquely outwards, becoming broader, and is

Inserted, tendinous and fleshy, into the inferior transverse ridge of the os occipitis, and into part of the concavity above that ridge.

Situation : This muscle is situated obliquely between the occiput and the second vertebra of the neck. It lies under the complexus; its outer fibres also pass under the insertion of the obliquus capitis superior.

Use : To draw the head backwards, and to rotate it.

19. The RECTUS CAPITIS POSTICUS MINOR—*Arises,* tendinous and narrow, from an eminence in the middle of the back part of the atlas, or first cervical vertebra. It becomes broader, and is

Inserted, fleshy, into the inferior transverse ridge of the os occipitis, and into the rough surface betwixt that ridge and the foramen magnum.

Situation : It is partly covered by the rectus capitis posticus major.; but a portion of this pair of muscles is seen projecting between the recti majores, and is situated beneath the complexus.

Use.: To draw the head backwards.

20. OBLIQUUS CAPITIS SUPERIOR—*Arises*, tendinous, from the upper and posterior part of the transverse process of the first cervical vertebra.

Inserted, tendinous and fleshy, into the inferior transverse ridge of the os occipitis behind the mastoid process, and into a small part of the surface above and below that ridge.

Situation : This muscle is situated laterally between the occiput and atlas. It is inserted under the complexus and trachelo-mastoideus, and it covers some of the outer fibres of the insertion of the rectus capitis posticus major.

Use : To draw the head backwards.

21. OBLIQUUS CAPITIS INFERIOR—*Arises*, tendinous and fleshy, from the side of the spinous process of the dentata or second cervical vertebra. It forms a thick belly, and is

Inserted into the under and back part of the transverse process of the atlas.

Situation : This muscle is obliquely situated between the two first vertebræ of the neck, being covered by the complexus and trachelo-mastoideus. Its origin lies between the origin of the rectus capitis posticus major, and the superior insertion of the semi-spinalis colli.

Use : To rotate the head, by turning the first vertebra upon the second.

22. The MULTIFIDUS SPINÆ.

On removing the muscles of the spine which have been described, we find situated beneath them the Multifidus Spinæ. It is that mass of muscular flesh, which lies close to the spinous and transverse processes of the vertebræ, extending from the dentata to the os sacrum. The bundles of which it is composed seem to pass from the transverse, to be inserted into the spinous processes. Considered as one muscle, it

Arises, tendinous and fleshy, from the spinous processes and back part of the os sacrum, and from the posterior adjoining part of the os ilium; from the oblique and transverse processes of all the lumbar vertebræ; from the transverse processes of all the dorsal vertebræ; and from those of the cervical vertebræ, excepting the three first. The fibres arising from this extensive origin pass obliquely, to be

Inserted, by distinct tendons, into the spinous processes of all the vertebræ of the loins and back, and into those of the six inferior vertebræ of the neck. The fibres arising from each vertebra are inserted into the second or third spinous process above, and sometimes even higher.

Use : To extend the spine obliquely, or move it to one side. When both muscles act, they extend the vertebræ backwards, and keep the body erect.

The small muscles situated between the processes of the vertebræ are,

1. INTERSPINALES colli, dorsi, et lumborum.—These are small bundles of fibres, which fill up the spaces between the spinous processes of the vertebræ. Each of these little muscles arises from the surface of one spinous process, and is inserted into the next spinous process.

In the neck they are large, and appear double, as the spinous processes of the cervical vertebræ are bifurcated. In the back and loins they are indistinct, and are rather small tendons than muscles.

Use: To assist in keeping the spine erect.

2. The INTERTRANSVERSALES colli, dorsi, et lumborum, are small muscles which fill up, in a similar manner, the space between the transverse processes of the vertebræ. In the neck they are bifurcated and distinct, and in the loins they are strong and fleshy; but in the back they are slender or wanting.

Use: To bend the spine laterally.

CHAPTER XIII.

DISSECTION OF THE MUSCLES SITUATED BETWEEN THE RIBS, AND ON THE INNER SURFACE OF THE STERNUM.

THE muscles which fill up the space between the ribs are named Intercostals; they are disposed on each side of the thorax in two layers, and each layer consists of eleven muscles.

The INTERCOSTALES EXTERNI—*Arise* from the inferior acute edge of each superior rib, extending from the spine to near the junction of the ribs with their cartilages. The fibres run obliquely forwards and downwards, and are

Inserted into the upper obtuse edge of each inferior rib, from the spine to near the cartilage of the rib.

Situation: These muscles are seen, on removing the muscles which cover the thorax.

The LEVATORES COSTARUM are twelve small muscles situated on each side of the dorsal vertebræ. They are portions of the external intercostals. Each of these small muscles *arises* from the transverse process of one of the dorsal vertebræ, and passes downwards, becoming broader, to be inserted into the posterior surface and upper edge of the rib next below the vertebra, near its tubercle.

The first of these muscles passes from the last cervical vertebra, the eleven others from the eleven superior dorsal vertebræ. The three or four inferior Levatores are longer, and run down to the second rib below the transverse process from which they arise. Hence Albinus names them the Levatores Costarum Longiores et Breviores.

2. The INTERCOSTALES INTERNI—*Arise* from the inferior acute edge of each superior rib, beginning at the sternum, and extending as far as the angle of the rib. The fibres run obliquely downwards and backwards, and are

Inserted into the superior obtuse edge of each inferior rib and cartilage, from the sternum to the angle. Portions of the internal intercostals pass over one rib, and are inserted into the next below it.

Thus the intercostal muscles decussate, and are double on the sides of the thorax; but, from the spine to the angles of the ribs, there are only the external intercostals; and, from the cartilages to the sternum, only the internal, and a dense fascia covering them. The whole of the internal intercostals, and the back part of the external, are lined by the pleura.

Use: The two layers of intercostal muscles bring the ribs nearer to each other, and as the lower ribs are the most moveable, their usual action is to elevate the ribs, so as to enlarge the cavity of the chest. Some regard the internal intercostals, as depressors and expiratory muscles.

One pair of muscles is situated on the inner surface of the sternum.

The TRIANGULARIS STERNI, or *Sterno-Costalis*—

Arises, tendinous and fleshy, from the edge of the whole cartilago ensiformis, and also from the edge of the sternum, as high as the cartilage of the fourth rib. The fibres ascend obliquely upwards and outwards, and form a flat muscle, which is

Inserted, by three or four, triangular fleshy and tendinous terminations, into the cartilages of the second, third, fourth, fifth, and sixth ribs.

Situation : This muscle is subject to many variations; it lies on the inside of the cartilages of the ribs and sternum, and is lined by the pleura. It is often continued, by tendinous or fleshy slips, with the upper part of the transversalis abdominis.

Use : To depress the cartilages and the bony extremities of the ribs, and consequently to assist in lessening the cavity of the thorax.

CHAPTER XIV.

DISSECTION OF THE MUSCLES SITUATED ON THE ANTERIOR PART OF THE NECK CLOSE TO THE VERTEBRÆ.

Four pairs of muscles are here situated.

1. The Longus Colli—*Arises,* tendinous and fleshy, from the sides of the bodies of the three superior dorsal vertebræ, and from the anterior surface of the transverse processes of the third, fourth, fifth, and sixth cervical vertebræ.

Inserted, tendinous and fleshy, into the fore part of the bodies of all the vertebræ of the neck.

Situation : This long and thin muscle, broader below than above, lies on the fore part and sides of the vertebral column, from the third dorsal vertebra to the atlas, and is frequently described as consisting of two portions.

It lies behind the œsophagus, and behind the great vessels and nerves of the neck.

Use : To bend the neck forwards, and to one side.

2. The RECTUS CAPITIS ANTICUS MAJOR—*Arises,* tendinous and fleshy, from the anterior tubercles of the transverse processes of the third, fourth, fifth, and sixth cervical vertebræ. It forms a considerable fleshy belly, ascending obliquely inwards upon the bodies of the vertebræ.

Inserted into the basilar process of the os occipitis a little before the condyloid process.

Situation : This muscle lies behind the pharnyx and great vessels of the neck, and more outwardly than the longus colli, over part of which it passes.

Use : To bend the head forwards.

3. The RECTUS CAPITIS ANTICUS MINOR—*Arises,* fleshy, from the fore part of the body of the first vertebra of the neck, near its transverse process; and ascending obliquely, is

Inserted near the root of the condyloid process of the occipital bone, under the last-described muscle.

Situation : It is placed beneath the last muscle, but is much shorter and narrower.

Use : To bend the head forwards.

4. The RECTUS CAPITIS LATERALIS—*Arises,* fleshy, from the superior and front part of the transverse process of the atlas.

Inserted, tendinous and fleshy, into a scabrous ridge of the os occipitis, which extends from the condyloid process of that bone towards the mastoid process.

Situation : This short muscle, the smallest in the region, is situated immediately behind the internal jugular vein, where it comes out from the cranium.

Use : To incline the head to one side.

CHAPTER XV.

DISSECTION OF THE SUPERIOR EXTREMITY.

SECTION I.

OF THE SHOULDER AND ARM.

§ 1. OF THE FASCIA, CUTANEOUS VEINS AND NERVES.

In the thigh we saw a strong fascia, arising from the neighbouring bones and ligaments, firmly investing and supporting the muscles: but, on removing the integuments from the shoulder and arm, we do not meet with a fascia of the same regular appearance. The muscles of the upper extremity are, however, covered by a tendinous expansion, which in the arm is thin and weak, but, below the bend of the elbow, becomes dense and strong. This BRACHIAL APONEUROSIS takes its origin, in part, from the cellular tissue of the axilla, and from the tendons of the latissimus dorsi and pectoralis major, on each side of that cavity; also from the tendinous insertion of the deltoid, but the belly of that muscle is only covered by close cellular tissue;—at the back part of the arm, it is derived from the spine of the scapula, and fascia covering the infra-spinatus muscle. From this origin, it extends as a thin, transparent, and, in many places, cellular expansion, over the whole arm, enveloping the muscles, and also covering the brachial vessels and nerves, where they descend along the inside of the limb. At the lower part of the arm, it adheres to the *external* and *internal intermuscular ligaments*, is fixed to the condyles of the os humeri, and then passes on to form the strong fascia of the fore-arm.

In removing the integuments, we meet with several cutaneous veins and nerves.

The cutaneous veins * of the upper extremity are the following; but, below the elbow, they vary much in their arrangement.

1. The Basilic Vein is seen arising from a small vein on the outside of the little finger, named Salvatella. It then runs upwards, along the inside of the fore-arm, near the ulna, usually consisting of two chief branches, on the anterior and posterior surface of the fascia, which have been termed the *Anterior* and *Posterior Ulnar* or *Cubital Veins,* and of which the latter is commonly of much greater volume. It passes over the fold of the arm, near the inner condyle, and is here joined by the *Median Basilic.* It then continues to ascend on the inside of the arm, but soon becomes more deeply seated, piercing the brachial aponeurosis, and lying along the brachial artery, from which it is separated only by a layer of its sheath. As it approaches the neck of the humerus, it sinks deep betwixt the folds of the arm-pit, and terminates in=the axillary vein, which may be considered as a continuation of the basilic vein. It communicates with the cephalic vein, but receives few secondary veins above the elbow.

2. The Cephalic Vein begins on the back of the hand, by a plexus of veins, which unite into one trunk, always found lying between the thumb and metacarpal bone of the fore-finger, and named Cephalica Pollicis. It is seen ascending along the radius, as the *Superficial Radial Vein,* receiving cutaneous branches from both surfaces of the fore-arm. It passes over the bend of the arm near the external condyle, receiving at this point

* The veins are described from their origin in the fore-arm, for the sake of perspicuity ; they ramify above the fascia of the fore-arm.

the *median cephalic*, and then continues to ascend along the outside of the arm, near the outer border of the biceps flexor cubiti. It passes betwixt the edges of the deltoid and pectoral muscles, and then dips inwards, to enter the axillary vein, immediately under the clavicle.

3. The MEDIAN VEIN. Several veins are seen running along the middle of the anterior part of the fore-arm. The trunk formed by these veins is called the Mediana Major. It ascends on the flat part of the fore-arm, betwixt the basilic and cephalic veins, and bifurcates at the fold of the arm into two branches; 1. The *Mediana Basilica*, passing off obliquely to join the basilic vein; 2. The *Mediana Cephalica*, which joins the cephalic. A third or deep branch also passes off to join the deep-seated veins, Mediana Profunda.

Turning back the integuments from the axilla and inside of the arm, you observe several Cutaneous Nerves ramifying above the muscles; they consist of,

1. The *Internal Cutaneous nerve*, a branch of the axillary plexus. It is seen accompanying the basilic vein, twisting over, and sometimes beneath it, and descending superficially along the inside of the arm: its distribution will be described with the nerves of the superior extremity.

2. Another *Cutaneous nerve*, frequently termed the *Nerve of Wrisberg.**, is usually found descending on the inside of the arm, towards the inner condyle, resting on the fibres of the triceps extensor. This nerve commonly arises from the lower part of the axillary plexus, or it is a branch of the ulnar nerve, given off high up in the axilla.

---

* *N. cutaneus minor internus* of Klint and Wrisberg:—*Cutaneus internus* of Soëmmering; sometimes confounded with the intercostal filaments.

3. The upper part of the arm also receives two or three cutaneous nerves from the dorsal or intercostal nerves, which come out of the thorax between the ribs. One of these nerves, derived from the anterior branch of the second dorsal, is often prolonged along the inner and back part of the arm, and is lost about the elbow.

4. The shoulder and back part of the scapula receive twigs from the cervical nerves.

5. The external cutaneous, ulnar, and spiral nerves, also send twigs to the integuments of the arm and fore-arm.

One or two small lymphatic glands are generally found on the inside of the arm above the inner condyle, placed superficially.

§ 2. MUSCLES SITUATED ON THE SHOULDER AND ARM.

These are ten in number, and may be divided into two classes;—the muscles situated on the shoulder, and the muscles situated on the arm.

In dissecting these muscles, little further precaution is necessary, than to take off the integuments and fascia in the direction of their fibres, and to preserve the principal vessels and nerves. If the arm has been detached, a block is to be placed under the shoulder, to make the fibres of the deltoid tense.

MUSCLES situated on the SHOULDER.

1. The DELTOIDES—*Arises*, tendinous and fleshy, from the external or posterior third of the clavicle, from the whole front edge of the acromion, and from the lower margin of the whole spine of the scapula. From these several origins, the fibres run downwards in different directions, and converge. Those arising from the clavicle run outwards and downwards: those from the spine of the scapula obliquely forwards, and downwards; and those from the acromion directly downwards.

Inserted, tendinous, into a triangular rough surface on the outer side of the os humeri, near its middle.

Situation: This muscle is entirely superficial, except where the thin fibres of the platysma myoides arise from its anterior surface. It arises from the same extent of bone as the trapezius is inserted into, that muscle passing upwards, while the deltoid runs downwards. It forms a strong coarse muscle, consisting of large fasciculi of fibres, and is separated, anteriorly, from the pectoralis major, by a cellular interval and the cephalic vein. It conceals the insertion of the pectoralis major, and the origins of the biceps flexor cubiti and coraco-bachialis, and covers the whole of the fore part and outside of the shoulder-joint, giving it the rounded appearance. Its insertion is situated betwixt the biceps flexor cubiti and the short head of the triceps extensor, and immediately above the origin of the brachialis internus. Its external surface is quite fleshy; but, on cutting it across, its internal surface is found tendinous: this surface is connected to the parts beneath by loose cellular tissue, and where it slides over the great tuberosity of the humerus, there is a *large bursa*, extending upwards under the acromion.

From the insertion of the deltoid to the outer condyle of the os humeri, is extended an Intermuscular Ligament, which separates the muscles on the anterior part of the arm from those on the posterior part, and gives attachment to the fibres of both. This may be considered as a tendinous septum, sent inwards from the brachial aponeurosis, to the lateral ridge of the humerus; it is named the EXTERNAL INTERMUSCULAR LIGAMENT.

Use: To raise the arm, and to move it forwards or backwards, according to the direction of its fibres.

Detach the deltoid from the scapula and clavicle, and

turn it downwards on the arm, that you may expose more completely the muscles on the dorsum of the former bone.

The following two muscles, which fill up the posterior surface of the scapula, above and below its spine, are covered by a *strong fascia*, which adheres to the spine and edges of that bone. On dissecting off this fascia, the fleshy fibres of the muscles will be found arising from its inner surface.

2. The SUPRA SPINATUS—*Arises*, fleshy, from all that part of the base of the scapula that is above its spine, from the superior costa as far forwards as the semilunar notch, from the spine itself, and from the concave surface betwixt it and the superior costa. The fleshy fibres, as they approach the neck of the scapula, terminate in a tendon, which passes under the acromion, slides over the neck of the scapula, (to which it is connected by loose cellular membrane,) adheres to the capsular ligament of the shoulder-joint, and is

Inserted into the anterior and superior part of the great tuberosity near the head of the os humeri.

Situation: This muscle fills up the fossa or cavity above the spine of the scapula, and is entirely concealed. Its belly is loosely covered by the fibres of the trapezius passing into the spine of the scapula; and its tendon passes under the deltoid:—at the upper part of the muscle, near the notch of the scapula, we see the origin of the omo-hyoideus.

Use: To raise the arm.

3. The INFRA-SPINATUS—*Arises*, principally fleshy, from the lower part of the spine of the scapula, as far back as the triangular flat surface; from the base of the bone below the spine to near the inferior angle; from the posterior ridge of the inferior costa; and from all the

dorsum of the bone below the spine. The fibres ascend and descend, towards a middle tendon, which runs forwards over the neck of the bone, and adheres to the capsular ligament.

Inserted, by this strong short tendon, into the middle part of the great tuberosity of the os humeri.

Situation : This muscle is in part concealed. The anterior part of its belly, and its tendinous insertion, are covered by the deltoid, and the trapezius passes over its upper and back part, but a considerable portion of the belly of the infra-spinatus is seen betwixt these two muscles, and above the superior fibres of the latissimus dorsi. It is inserted below the tendon of the supra-spinatus, the two tendons being in general united.

Use : To roll the humerus outwards, to assist in raising the arm, and in moving it outwards when raised.

The next muscle, which is small and narrow, ranges along the inferior border of the infra-spinatus.

4. The TERES MINOR—*Arises*, fleshy, from the narrow depression between the two ridges in the inferior costa of the scapula, extending from the neck of the bone to within an inch or two of the inferior angle. It passes forwards along the inferior edge of the infra-spinatus, adheres to the capsular ligament of the shoulder-joint, and is

Inserted, tendinous and fleshy, into the lower and back part of the great tuberosity of the os humeri.

Situation : It is inserted below the tendon of the infra-spinatus. Its origin lies between the infra-spinatus and teres major, and is partly concealed by them. The middle of its belly is superficial, and not covered by any muscle ; but its insertion is concealed by the deltoid. The fascia which covers the infra-spinatus, envelopes also the teres minor : and the two muscles are, in some subjects, so closely united, as to be with difficulty separated.

Use: To draw the humerus downwards and backwards, and to rotate it outwards.

On the lower edge of the teres minor, we see a distinct and larger muscle.

5. The TERES MAJOR—*Arises,* from an oblong, rough, flattened surface, at the inferior angle of the scapula. It forms a thick belly, which passes forwards and upwards towards the inside of the arm.

Inserted, by a broad thin tendon, into the inner or posterior edge of the bicipital groove of the humerus.

Situation: The origin of this muscle is superficial, adhering to the lower fibres of the infra-spinatus; or it is partially overlapped by the latissimus dorsi, which crosses it to reach the axilla. The belly passes before the long head of the triceps extensor cubiti. The tendon is inserted along with the tendon of the latissimus dorsi, assisting that muscle to form the posterior fold of the axilla. Observe the relative situation of these tendons; the tendon of the latissimus dorsi is anterior to the tendon of the teres major, but the lower edge of the latter extends further down the arm; they both pass under the coraco-brachialis and short head of the biceps flexor, to reach the place of their insertion, at first appearing inseparably united, but, on dividing them with some care, we find an intermediate cavity or bursa. A narrow *tendinous band* extends from the lesser tuberosity of the humerus, arching over these tendons, and binding them down, and giving insertion to some of the fibres of the coraco-brachialis.

Use: To rotate the humerus inwards, and to draw it backwards and downwards.

The next muscle occupies the fossa or inner surface of the scapula.

6. The SUBSCAPULARIS—*Arises*, fleshy, from all the base of the scapula internally, from the superior and inferior costæ, and from the whole internal surface of the bone. It forms a broad triangular muscle, consisting of several large tendinous and fleshy fasciculi, which converge, slide over the inner surface of the neck of the scapula, pass in the hollow under the root of the coracoid process, and adhere to the inner part of the capsular ligament of the shoulder-joint.

Inserted, by a strong tendon, into the lesser tuberosity near the head of the os humeri.

Situation: The whole of this muscle is concealed by the scapula and muscles of the shoulder. It lies betwixt that bone and the serratus magnus, being separated by loose cellular tissue from the serratus, and forming, with that muscle, the posterior hollow of the axilla. The tendon passes under the coraco-brachialis and short head of the biceps flexor, to reach the lesser tuberosity. The lower edge of this muscle is in contact with the upper edge of the teres major.

Use: To rotate the os humeri inwards, and to draw it to the side of the body.

Beneath the tendon of the subscapularis, at the neck of the scapula, there is a large bursa, which is generally connected with the capsule of the shoulder-joint by a wide opening.

MUSCLES situated on the arm.

7. The BICEPS FLEXOR CUBITI—*Arises* by two heads. The first and outermost, called the *Long Head*, arises, by a strong flattened tendon, from a smooth surface in the upper edge of the glenoid cavity of the scapula. It

passes over the head of the os humeri, within the capsular ligament of the shoulder-joint, and enters a groove betwixt the two tuberosities of that bone. It forms a strong fleshy belly. The second and innermost, called the *Short Head*, arises, tendinous, from the lower part of the coracoid process of the scapula, in common with the coraco-brachialis, and sends off a fleshy belly.

These two fleshy bellies are at first only connected by cellular membrane: they form a thick mass; and below the middle of the arm, become inseparably united. The single muscle thus formed descends in front of the arm, and, contracting in volume, terminates near the elbow-joint in a strong tendon, which is at first broad, but soon becomes rounder, and passes over the fore part of the joint. It then slides over the cartilaginous middle surface of the tubercle at the upper end of the radius, and is

Inserted into the posterior and internal rough part of that tubercle. A bursa mucosa is placed between the tendon and front of the tubercle.

Situation: The tendon of the long head cannot be seen till the capsular ligament of the shoulder is opened: it is found to be on the outside of the synovial bag of the joint, and it is invested by a kind of sheath, which is prolonged from the synovial membrane over the tendon. Where it runs in the bicipital groove of the os humeri, it lies betwixt the pectoralis major and latissimus dorsi, and is firmly bound down by fibres passing from the capsular ligament and adjacent tendons. The short head arises from the coracoid process, betwixt the origin of the coraco-brachialis, and the strong ligament which passes from the coracoid process to the acromion. These two origins are concealed by the deltoides and pectoralis major. The belly of the muscle is immediately under the integuments, and so is the tendon where it passes over the elbow-joint. It is seen sinking between the supinator radii longus and pronator teres, to arrive at its point of insertion, having previously, at the bend of the elbow, sent off from its inside an APONEU-ROSIS, which assists in forming the fascia of the fore-arm.

Use: To turn the hand supine, to bend the fore-arm on the arm, and the arm on the shoulder; and to brace the fascia of the fore-arm.

8. The CORACO BRACHIALIS—*Arises,* tendinous and fleshy, from the middle part of the apex of the coracoid process of the scapula. Its fibres, as it descends, also arise from the edge of the short tendon of the biceps flexor cubiti. It forms a flat fleshy belly, which is usually perforated by the nerve, named Musculo-Cutaneus.

Inserted, tendinous and fleshy, about the middle of the internal part of the os humeri, into a rough ridge.

Situation: This muscle is much connected with the short head of the biceps flexor cubiti. It arises betwixt that muscle and the origin of the pectoralis minor. In the arm, it lies behind and on the inside of the biceps, and is concealed by the pectoralis major and deltoides, excepting a small part of it which is seen projecting betwixt the biceps flexor and triceps extensor cubiti. It is inserted immediately below the tendons of the latissimus dorsi and teres major, and before the brachialis externus. The lower part of its insertion passes betwixt the brachialis internus and brachialis externus.

The INTERNAL INTERMUSCULAR LIGAMENT is seen extending from the lower part of this muscle along a ridge to the internal condyle, passing inwards from the fascia of the arm, and separating the brachialis internus from the brachialis externus, or third head of the triceps extensor cubiti.

Use : To move the arm upwards and forwards.

9. The BRACHIALIS INTERNUS—*Arises,* from the middle of the os humeri, by two fleshy slips, which pass on each side of the insertion of the deltoid muscle ;— and fleshy from all the fore part of the bone below, nearly as far as the condyles, being connected also with the intermuscular ligaments. The fibres converge, pass over the front of the elbow-joint, and adhere loosely to its capsule and anterior ligament.

Inserted, by a strong short tendon, into the rough surface immediately below the coronoid process of the ulna.

Situation : The most external of the fleshy slips of this muscle lies between the deltoid and short head of the triceps extensor, the internal between the deltoid and coraco-brachialis. The belly is almost entirely concealed by the biceps flexor cubiti, excepting a small portion which projects beyond the outer edge of that muscle. The tendon dips down betwixt the supinator radii longus and pronator teres, crosses under the tendon of the biceps flexor, and is inserted on the inside of that tendon.

Use : To bend the fore-arm.

10. The TRICEPS EXTENSOR CUBITI is the great muscle which covers all the back part of the arm. It *arises* by three heads. The first, or LONG HEAD, *arises,* by a short, flat tendon, from the inferior costa of the scapula near its neck, and forms a large belly, which covers the back part of the os humeri. The second, or SHORT HEAD, *arises,* on the outer and back part of the os humeri, by an acute tendinous and fleshy beginning, from a ridge which runs from the back part of the great tu-

berosity towards the outer condyle. The fibres begin to arise a little below the tuberosity, and are continued down to the condyle; they also arise from the surface of bone behind the ridge, and from the intermuscular ligament which separates them from the muscles on the fore part of the arm. The third head, called BRACHIALIS EXTERNUS, *arises,* by an acute beginning, from the inside of the os humeri above its middle, and from a ridge extending to the inner condyle; it also arises from the surface behind this ridge, and from the internal intermuscular ligament.

The three heads unite above the middle of the os humeri, and adhere to the whole back part of the bone, receiving fleshy fibres down to the posterior fossa. They form a thick strong tendon, which is

Inserted into the rough back part of the process of the ulna, called Olecranon, and partly into the condyles of the os humeri, adhering by cellular tissue to the capsule and ligaments of the elbow-joint.

Situation: The long head, where it arises from the scapula, is concealed by the deltoid; it arises betwixt the teres minor and teres major, and passes betwixt those muscles to the arm, and, in the remainder of its course, is superficially seated. The short head arises immediately below the insertion of the teres minor, its upper part is therefore covered by the deltoides: below the deltoid, it is superficial, and arises more outwardly than the brachialis internus, supinator radii longus, and radial extensors of the carpus. The brachialis externus is situated immediately under the integuments; it begins to arise below the insertion of the latissimus dorsi and teres major, and passes down the arm between the triceps longus and coraco-brachialis at first, afterwards between the triceps longus and brachialis internus. The tendon of the triceps sends off a thin fascia, which co-. vers the triangular surface of the ulna, on which we commonly lean. Numerous fibres are also sent off, to. assist in forming the fascia of the fore-arm. There is a bursa between the tendon and the olecranon, and a larger *subcutaneous* bursa is interposed between the skin and aponeurotic fibres covering that process.

Use : To extend the fore-arm. The long head will also assist in drawing the arm backwards.

SECTION II.

DISSECTION OF THE FASCIA AND MUSCLES SITUATED ON THE CUBIT OR FORE-ARM *.

§. 1. ÓF THE FASCIA AND OF THE PARTS AT THE BEND OF THE ARM.

ON removing the integuments of the fore-arm, we find, as in the leg, a strong fascia investing all the muscles. This fascia is generally thick and strong, but especially so on the back part of the limb : it is evidently conti- nued from the fascia of the arm, and from the intermus- cular ligaments, which pass down to the condyles of the os humeri. It is attached to the condyles, and it ad- heres firmly to the olecranon of the ulna. It receives, on the posterior part, a great addition of fibres from the tendon of the triceps extensor ; and on the fore part of the arm, it appears to be a continuation of the aponeu- rosis which is sent off from the biceps flexor cubiti. It descends over the fore-arm, binding down the muscles, and sending processes between them. On the outer

* In the following description, the palm of the hand is supposed to be turned forwards, so that the radius and thumb are upon the outer side of the fore-arm, and the ulna and little finger upon its inner side.

Q

border of the arm, it is unattached, but, on the inside, it is fixed to the ulna nearly in its whole length, and it is continued, below, into the Annular Ligaments of the wrist.

Above the fascia, we meet with several cutaneous veins and nerves. The Veins have been already described. The Nerves are chiefly branches of the Internal and External Cutaneous Nerves, and of the cutaneous branch of the Radial nerve. They are usually reflected off, adhering to the integuments, but may easily be traced.

The relative situation of the Vessels at the bend of the arm should be well attended to. The cutaneous veins here situated, vary much in size, and the cutaneous nerves, in passing over the fold of the arm, are connected with the veins, sending some filaments in front and others behind them. The Vena basilica is seen passing over the bend of the arm near the inner condyle, the Cephalica situated near the outer condyle,. and each of these veins receives a branch passing obliquely from the Vena mediana major. These vessels lie above the fascia, while the Brachial Artery lies deeper beneath the fascia, in a hollow having some resemblance to the axilla. The Artery is seen descending over the joint near the inner condyle, on the inside of the tendon of the biceps flexor cubiti, and then sinking beneath the aponeurosis, which is sent off from this tendon to the common fascia of the fore-arm. Divide this aponeurosis, and you find the artery embedded in cellular substance, and lodged in the triangular space, which is here formed between the pronator teres and flexor muscles of the wrist and fingers on one side, and the supinator longus and extensors on the other: it is found close to the inner side of the tendon of the biceps, resting on the fibres of the brachialis internus; it is accompanied by two veins, and on its

inner side, or nearer the condyle, is the Median nerve.
In this hollow, it divides into the Radial and Ulnar arteries, which latter gives off the Interosseous.

Of the cutaneous veins, the Median Basilic is more
immediately in front of the artery, ascending in a line
nearly parallel to it, at first separated from the artery
by the interposed aponeurosis of the biceps, but, higher
up, in close proximity before it.

The fascia of the fore-arm is now to be dissected off
the muscles. Like the fascia of the leg, it is found
firmly attached to the muscular fibres, so that the exposed surface appears ragged : and septa, or partitions,
named Intermuscular Ligaments or tendons, are seen
proceeding inwards to connect the muscles, and to give
origin to many of their fibres.

§. 2. MUSCLES SITUATED ON THE FORE PART OF THE CU-
BIT, AND ARISING FROM THE INNER CONDYLE OF THE
OS HUMERI.

These are eight in number, and may be divided into
two classes; the superficial, and the deep-seated.

1. THE SUPERFICIAL.

All the muscles passing from the inner condyle, may
be said to arise by one common tendinous head from
the condyle; and this head may be said to divide into
the different muscles; but they will be here described
as arising distinct from the condyle. It must, however,
be recollected, that their origins are intimately connected by intermuscular ligaments, and that they cannot be
separated without dividing some of their fibres.

1. The PRONATOR RADII TERES—*Arises*, tendinous
and fleshy, from the anterior surface of the inner condyle of the os humeri, and, by a small distinct tendon,
from the coronoid process of the ulna. It also arises

from the fascia of the fore-arm, and from the tendinous partitions, which separate it from the flexor radialis, and flexor sublimis. The fibres pass outwards, run by the side of the tubercle of the radius, and pass over the outer edge of that bone, to be

Inserted, tendinous and fleshy, into a rough surface on the back part of the radius, about its middle.

Situation : Of the muscles which pass from the internal condyle, the pronator teres is situated nearest the outer edge of the arm. It is in great part superficial, and lies immediately on the inside of the tendon of the brachialis internus, being separated from the supinator longus by the triangular hollow, in which the biceps tendon is lodged, with the brachial artery and median nerve: that nerve passes betwixt the two origins of the pronator. The tendon, to arrive at its place of insertion, passes under the belly of the supinator longus, and, when that muscle is removed, will be found inserted into the radius immediately below the supinator brevis.

Use : To roll the radius, together with the hand, inwards, producing the state of pronation: to bend the fore-arm.

2. The FLEXOR CARPI RADIALIS, or RADIALIS INTERNUS—*Arises,* by a narrow tendinous beginning, from the lower and fore part of the internal condyle of the os humeri: fleshy from the fascia and intermuscular ligaments, and from the upper end of the ulna. It forms a thick belly, which runs down the fore-arm, and terminates in a flat tendon. This tendon passes over the carpus, in a distinct fibrous canal, formed by the annular ligament * of the wrist and a deep groove in the os trapezium, and is

* The ANNULAR LIGAMENT of the wrist consists of two parts. 1. The *ligamentum carpi dorsale vel posterius* passes from the styloid

Inserted into the fore part of the base of the metacarpal bone sustaining the fore-finger.

Situation: This muscle is situated immediately under the fascia, excepting its upper extremity, over which the pronator teres crosses. It arises between the pronator teres and palmaris longus, and descends betwixt those muscles. Below the insertion of the pronator, it is situated betwixt the supinator radii longus and palmaris longus. The strong tendon is prominent in the wrist, passing towards the ball of the thumb: its insertion cannot be seen till the palm of the hand is dissected, where it will be found concealed by the short muscles of the thumb: and there is a synovial bursa surrounding the tendon, where it traverses the trapezium.

Use: To bend the hand, and to assist in its pronation.

3. The PALMARIS LONGUS—*Arises*, by a slender tendon, from the fore part of the inner condyle of the os

process of the ulna and os pisiforme, transversely, over the back of the wrist, and spreads out broad, to be affixed to the styloid process of the radius. It appears to be a portion of the fascia of the fore-arm, strengthened by additional transverse fibres: under it pass the Extensor tendons, contained in separate sheaths or canals, which are formed by the attachment of these dorsal ligamentous fibres to the projecting ridges on the back part of the radius and ulna.

2. The *Ligamentum carpi internum,* or *anterior annular ligament* is much stronger, and passes across the fore part of the wrist. It arises from the os pisiforme and os unciforme on the inner edge of the wrist, and is attached to the os scaphoides and os trapezium on the outer edge: the upper border is continuous with the fascia of the fore-arm, and its lower edge is intermixed with the Palmar fascia. This anterior annular ligament forms, with the carpal bones, a deep canal, to give passage to the tendons of the common flexors of the fingers, and of the long flexor of the thumb; but the flexor ulnaris and palmaris tendons are external to it, and the tendon of the flexor radialis only sinks under that portion of the ligament, which springs from the trapezium.

humeri; and fleshy, from the intermuscular ligaments; it forms a short fleshy belly, which soon sends off a long slender tendon. This tendon descends along the fore-arm, and is

Inserted, near the root of the thumb, into the ante-rior annular ligament, but its chief portion expands into the broad tendinous membrane that covers the palm of the hand, named Fascia, or Aponeurosis Palmaris.

Situation : It arises betwixt the flexor carpi radialis and flexor ulnaris. Its tendon descends betwixt these two muscles, and above the fibres of the flexor digito-rum sublimis. This muscle is sometimes wanting.

Use : To bend the hand on the fore-arm, and to stretch or tighten the palmar fascia.

4. The FLEXOR CARPI ULNARIS, or ULNARIS INTER-NUS—*Arises*, tendinous, from the inferior part of the in-ternal condyle of the os humeri; tendinous and fleshy, from the inner side of the olecranon, and by a tendinous expansion from the posterior ridge * of the ulna, to near the lower end of the bone. It also arises from a short intermuscular septum, which separates it above from the flexor sublimis, and from the fascia of the fore-arm. The fibres pass obliquely forwards into a tendon, which runs over the fore part of the ulna, and is

Inserted into the os pisiforme, sending off some fibres over the short muscles of the little finger. The tendon is also bound down by a thin aponeurotic slip, which passes from the upper part of the anterior annular liga-ment of the wrist, crossing over the ulnar artery and nerve.

Situation : This muscle arises behind the other mus-cles which pass from the internal condyle. It runs

---

* The ridge which leads to the styloid process.

superficially along the inner or ulnar edge of the fore-arm, between the flexor sublimis on the fore part, and the extensor carpi ulnaris on the back part of the ulna. Between the two origins from the condyle and olecranon, the ulnar nerve passes, being covered by an aponeurosis, which proceeds from the one portion of the muscle to the other. There is a bursa between the tendon and the os pisiforme.

Use: To bend the hand.

5. The FLEXOR SUBLIMIS PERFORATUS—*Arises*, tendinous and fleshy, from the under part of the internal condyle of the os humeri; tendinous, from the lower part of the coronoid process of the ulna; fleshy, from the tubercle of the radius, from the middle of the fore part of that bone, and from the middle third of its outer edge. These origins form a strong fleshy mass, which descends along the middle of the fore-arm, and sends off four tendons. The tendons are connected by cellular membrane, and pass together under the annular ligament of the wrist; after which they separate, become thinner and flatter, pass along the metacarpal bone and first phalanx of each of the four fingers, and are

Inserted into the anterior and upper part of the second phalanx, each tendon being divided, near the extremity of the first phalanx, for the passage of a tendon of the flexor profundus.

Situation: To expose the origin of this muscle, the bellies of the pronator teres, flexor carpi radialis and palmaris longus, must be detached from the condyle. It arises behind these muscles, and is connected to them by intermuscular ligaments. It descends along the fore-arm, under these muscles, but a part of it is seen projecting towards the inner edge of the arm, betwixt the tendons of the palmaris longus and flexor carpi ulnaris.

It arises from the radius immediately below the insertion of the biceps flexor cubiti, and on the inside of the insertion of the pronator teres. Its tendons will be seen in the dissection of the palm of the hand.

Use: To bend the second joint or phalanx of the fingers, and the hand on the fore-arm.

By removing the belly of the flexor sublimis, we expose,

2. THE DEEP-SEATED MUSCLES.

6. The FLEXOR PROFUNDUS PERFORANS — *Arises,* fleshy, from the smooth concavity on the inside of the ulna, betwixt the coronoid process and the olecranon; from the smooth flat surface of the ulna, betwixt its posterior and internal angles; from the under part of the coronoid process; from the fore part of the ulna below that process, and betwixt the internal angle and that angle which gives attachment to the interosseous ligament. It also arises from the inner half of the interosseous ligament. This muscle forms a thick mass, which descends along the fore part of the ulna, adhering to that bone as low as one-third of its length from its inferior extremity, and terminating in sending off four tendons. These tendons are flat, pass together under the annular ligament of the wrist, run through the slits in the tendons of the flexor sublimis, and are

Inserted into the anterior and upper part of the third or last phalanx of all the fingers.

Situation: This muscle lies deeply on the ulnar side of the fore-arm, concealed by the flexor sublimis and flexor carpi ulnaris. Its tendons will be seen in dissecting the hand.

Use: To bend the last joint of the fingers, and the hand on the fore-arm.

7. FLEXOR LONGUS POLLICIS MANÛS—*Arises,* by an

acute fleshy beginning, from the upper and fore part of the radius, immediately below its tubercle; fleshy, from the outer edge and anterior surface of that bone, as low as two inches above its inferior extremity, and from the outer part of the interosseous ligament. It has also generally a tendinous origin from the internal condyle of the os humeri. This origin forms a distinct fleshy slip, which is joined to the inner and upper part of the portion of the muscle arising from the radius. The fibres pass obliquely into a tendon on the anterior surface of the muscle. The tendon passes under the annular ligament of the wrist, with the flexor tendons of the fingers, runs between the two heads of the short flexor of the thumb, and between the two sesamoid bones, and is

Inserted into the base of the extreme phalanx of the thumb.

Situation: This muscle lies on the outer or radial side of the flexor profundus, resting on the anterior surface of the radius, and the anterior interosseous artery and nerve descend along the interstice, which separates the two muscles. That portion of the flexor pollicis, which arises from the inner condyle, passes over the belly of the flexor profundus, and under the flexor sublimis. The whole muscle is concealed by the flexor sublimis; its tendon will be seen in dissecting the short muscles of the thumb.

Use: To bend the last joint of the thumb, and the hand on the fore-arm.

On separating the lower part of the two last-described muscles, we expose a small square muscle, passing transversely just above the wrist.

8. The Pronator Quadratus—*Arises*, broad, tendinous, and fleshy, from the inner edge of the ulna, extending from the lower extremity of the bone two

inches up its edge. The fibres run transversely, adhere to the interosseous ligament, and are

Inserted into the lower and anterior part of the radius.

Situation : This muscle lies close to the bones, covered by the flexor longus pollicis and flexor digitorum sublimis.

Use : To turn the radius, together with the hand, inwards.

§. 3. MUSCLES SITUATED ON THE OUTER AND BACK PART OF THE FORE-ARM, AND ARISING FROM THE OUTER CONDYLE OF THE OS HUMERI.

These muscles are twelve in number, and may be divided into two classes : 1. The Superficial; and, 2. The Deep-seated.

1. THE SUPERFICIAL.

The muscles which arise from the outer condyle, are much more distinct in their origins, than those which arise from the inner condyle. Several of them arise a considerable way up the os humeri ; but there is here also a common tendinous origin, from which the extensor carpi radialis brevior, extensor digitorum communis, and extensor carpi radialis proceed; so that these muscles are intimately connected.

1. SUPINATOR RADII LONGUS—*Arises*, tendinous and fleshy, from the external ridge of the os humeri, which leads to the outer condyle. It begins to arise nearly as far up as the middle of the bone, just below the insertion of the deltoid, and ceases to adhere about two inches above the condyle. It forms a thick fleshy belly, which passes over the side of the elbow-joint, becomes smaller, and terminates above the middle of the fore-arm in a flat tendon. The tendon becomes gradually rounder, and is

Inserted into a rough surface on the outer side of the

inferior extremity of the radius, near the base of the styloid process.

Situation: This muscle is situated immediately under the integuments along the outer or radial edge of the arm and fore-arm, and it forms the outer boundary of the triangular space at the fold of the arm, in which the tendon of the biceps and the brachial artery are lodged. Its origin from the humerus lies betwixt the brachialis internus and short head of the triceps extensor cubiti, from which latter muscle it is separated by the external intermuscular ligament. It descends along the radius, placed at first between the extensor carpi radialis longior and pronator teres, lower down between the tendons of the flexor carpi radialis and extensor radialis longior. Its insertion is crossed by the extensors of the thumb.

Use: To roll the radius outwards, and turn the palm of the hand upwards, producing supination; also to bend the fore-arm on the humerus.

2. The EXTENSOR CARPI RADIALIS LONGIOR, or RADIALIS EXTERNUS LONGIOR—*Arises,* tendinous and fleshy, from the external ridge of the os humeri, beginning immediately below the origin of the supinator longus, and continuing to arise as far as the upper part of the outer condyle. It forms a thick short belly, which passes over the side of the elbow-joint, and terminates, above the middle of the radius, in a flat tendon. The tendon runs along the radius, and, becoming rounder, passes through a groove in the back part of the inferior extremity of that bone, and over the carpus, to be

Inserted into the posterior part of the base of the metacarpal bone of the fore-finger.

Situation: The belly lies under the supinator longus, but part of it projects behind that muscle. The tendon descends behind that of the supinator, and passes under the extensors of the thumb, and annular ligament of the wrist, to arrive at the place of its insertion : where the tendons of this and the next muscle pass through the groove of the radius, they are surrounded by a synovial sheath, which is prolonged nearly to their points of insertion.

Use : To extend the wrist, and move the hand backwards, and to assist in bending the fore-arm.

3. The EXTENSOR CARPI RADIALIS BREVIOR, or RADIALIS EXTERNUS BREVIOR—*Arises,* tendinous, from the under and back part of the external condyle of the os humeri, and from the external lateral ligament of the elbow-joint; and also from the intermuscular ligament connecting it with the extensor digitorum. Its thick belly runs along the outside of the radius, and terminates in a tendon, which passes through the same groove in the radius as the extensor radialis longior, and under the annular ligament, to be

Inserted into the upper and back part of the metacarpal bone that supports the middle finger.

Situation : This muscle lies partly under the extensor radialis longior, but it also projects behind it. The tendon is seen behind, or more inwardly than the tendon of the long extensor, and diverges from it to reach the middle finger; it passes under the extensors of the thumb and indicator.

Use : The same as that of the last muscle.

4. The EXTENSOR DIGITORUM COMMUNIS—*Arises,* tendinous, from the under part of the external condyle of the os humeri; fleshy from the intermuscular ligaments which connect it on one side to the extensor carpi radialis brevior, and on the other to the extensor minimi digiti, and from the inner surface of the fascia of the fore-arm. It descends along the back part of the fore-

arm, and adheres to the ulna where it passes over it. The fleshy belly terminates in·four flat tendons, which are connected by loose cellular tissue, pass under the annular ligament, in a depression on the back part of the radius, and then, separating and becoming broader, are continued over the metacarpal bones, to be

Inserted into the posterior part of all the phalanges of the four fingers by a tendinous expansion.

Situation : It arises betwixt the extensor radialis brevior and the extensor minimi digiti, descends. betwixt these.muscles, and is situated immediately under the fascia and integuments. The tendons are connected on the back of the metacarpal bones, by cross slips, and the three last are in general slit longitudinally: Where they pass under the annular ligament, they are invested by a bursa, which prolongs a synovial sheath over each of them.

. *Use :* To extend all the joints of the fingers, and bring them backwards.

The tendon of the common extensor going to the little finger is sometimes wanting : that finger always has a proper extensor, viz.

5. EXTENSOR PROPRIUS MINIMI DIGITI, or AURICULARIS—*Arises,* by a narrow origin, from the. external condyle of the os humeri, and from the intermuscular ligaments which connect it with the extensor communis and extensor carpi ulnaris ; and also from the fascia of the fore-arm. It forms an extremely slender fleshy belly, which descends and terminates in a tendon : the tendon passes through a separate depression of the radius, and a particular ring of the annular ligament, being frequently split into two contiguous portions, which reunite : it adheres to the edge of the fourth tendon of the extensor digitorum communis, and is

Inserted, in the same manner, into the posterior surface of the phalanges of the little finger.

Situation: It is placed between the extensor carpi ulnaris and extensor communis, and is frequently considered as a portion of the latter.

Use : To assist the fourth portion of the last muscle in extending the little finger.

The posterior surface of each finger is covered with a tendinous expansion, which is formed by the tendons of the common extensor, of the lumbricalis, and interossei, and which terminates in the third or extreme phalanx. This tendinous expansion will be found to split, near the first digital joint, into three portions, of which the middle one is inserted into the second phalanx, while the two lateral portions proceed forwards over the side of the joint, reunite, and form a flat tendon, which is inserted into the base of the third phalanx.

6. The. EXTENSOR CARPI ULNARIS, or ULNARIS EXTERNUS—*Arises*, tendinous, from the upper part of the external condyle ; fleshy, from the intermuscular ligaments and inside of the fascia. It crosses towards the ulna, and arises, fleshy, from the back part of that bone, and from the posterior ridge leading to the styloid process, in the middle third of its length. It terminates in a strong tendon, which passes through a groove in the back part of the lower end of the ulna, under the posterior annular ligament, and over the carpus, and, sinking under the abductor minimi digiti, is

Inserted, finally, into the posterior and upper part of the metacarpal bone of the little finger, sending off some aponeurotic fibres over its adductor muscle.

Situation: This muscle is entirely superficial. It arises from the condyle betwixt the extensor minimi digiti and anconeus. It descends along the fore-arm betwixt the extensor minimi digiti and flexor carpi ulnaris : and it has a fibrous sheath lined with synovial membrane, where it passes under the annular ligament, and over the ulna and bones of the carpus.

Use: To extend the wrist, and bring the hand backwards; but chiefly to bend the hand laterally towards the ulna.

7. The ANCONEUS is a small triangular muscle, situated at the outer side of the olecranon, covered by a strong aponeurosis, which is part of the general fascia of the fore-arm.

It *arises*, tendinous, from the posterior and lower part of the external condyle of the os humeri ; forms a thick triangular fleshy mass, adhering to the capsule and ligaments of the elbow-joint, and is

Inserted into the concave surface on the outer side of the olecranon, and into the posterior edge of the ulna for some distance.

Situation : This muscle lies betwixt the upper part of the extensor carpi ulnaris and the olecranon. Its upper edge is closely united with the triceps extensor cubiti, so that, in some subjects, the line of division is indistinct : the fascia covering it is very strong, and does not adhere to the muscle, and is more immediately derived from the tendon of the triceps.

Use: To assist in extending the fore-arm.

By removing the superficial muscles, we expose

2. THE DEEP-SEATED.

8. The SUPINATOR RADII BREVIS—*Arises*, tendinous, from the lower part of the external condyle of the os humeri ; tendinous and fleshy, from the ridge running down from the coronoid process along the posterior surface of the ulna. The fibres adhere firmly to the exter-

nal lateral ligament of the elbow-joint, pass outwards round the upper part of the radius, and are

Inserted into the upper and outer edge of the tubercle of the radius, and into an oblique ridge extending from the tubercle downwards and outwards to the insertion of the pronator teres.

Situation : This muscle nearly surrounds the upper and outer part of the radius. It is concealed at the outer edge of the arm by the supinator longus and extensores carpi radiales ; behind, by the extensor digitorum, extensor minimi digiti, extensor carpi ulnaris, and anconeus ; before, by the brachialis internus, and by the tendon of the biceps flexor cubiti, close to which tendon this muscle is inserted.

Use : To roll the radius outwards, and bring the hand supine.

On the back part of the fore-arm we meet with three muscles going to the thumb, and one to the fore-finger.

9. The EXTENSOR OSSIS METACARPI PÓLLICIS MANÛS, or ABDUCTOR LONGUS POLLICIS—*Arises*, fleshy, from the middle and posterior part of the ulna, immediately below the termination of the anconeus ; from the interosseous ligament, and from the posterior surface of the radius below the insertion of the supinator radii brevis. The fleshy fibres cross the fore-arm obliquely towards the root of the thumb, and terminate in a tendon, which passes through a groove in the outer edge of the lower extremity of the radius, and is confined by a separate ring of the annular ligament.

Inserted, generally by two tendons, into the os trapezium, and into the upper and back part of the metacarpal bone of the thumb.

Use : To extend the metacarpal bone of the thumb outwardly, separating it from the fingers.

10. The Extensor Primi Internodii-Pollicis Manûs, or Extensor Minor Pollicis,—is shorter and less thick than the last muscle. It *arises*, fleshy, from the ulna below its middle, and from the interosseous ligament, but chiefly from the posterior surface of the radius, lying below and on the ulnar side of the extensor ossis metacarpi: following the same direction, it forms a tendon, which passes through the same groove with the tendon of that muscle, (or sometimes in a narrow separate depression,) and is then continued over the metacarpal bone, to be

Inserted into the posterior part of the first phalanx of the thumb. Part of the tendon is also continued into the base of the second or extreme phalanx.

Use : To extend the first phalanx of the thumb obliquely outwards.

11. The Extensor Secúndi Internodii Pollicis Manûs, or Extensor Major Pollicis—is larger and longer than the last muscle. It *arises*, tendinous and fleshy, from the posterior surface of the ulna, above its middle, and from the interroseous ligament. Its belly partly covers the origins of the two other extensors of the thumb; it descends between the last muscle and the indicator, and terminates in a tendon, which runs under the annular ligament, apart from the two last tendons, through a distinct groove in the back part of the radius, and passes by the side of the metacarpal bone, and over the first phalanx of the thumb, to be

Inserted into the posterior and upper part of the second or extreme phalanx.

Use : To extend the last joint of the thumb obliquely backwards.

Situation of the extensors of the thumb.—The origins of these muscles are concealed by the extensor digitorum communis and extensor carpi ulnaris.—Their bellies are seen coming from betwixt the extensor digitorum communis and extensor carpi radialis brevior: the tendons cross over the tendons of the extensores carpi radiales, passing under the annular ligament of the wrist, to arrive at the place of their insertion, and, where they pass under that ligament, being surrounded by a sheath of synovial membrane. The tendons of the two first extensors are felt projecting, like a chord, from the styloid process of the radius to the root of the metacarpal bone of the thumb, while the third tendon or extensor major is separated from them by a considerable interval, in which are seen the terminations of the tendons of the extensores carpi radiales, and the radial artery. These tendons invest the back part of the thumb with a fascia.

12. The INDICATOR, or EXTENSOR INDICIS PROPRIUS —Arises, by an acute fleshy beginning, from the middle of the back part of the ulna, and from the adjacent interosseous ligament. Near the wrist, it ends in a tendon, which passes through the same sheath of the annular ligament with the tendons of the extensor digitorum communis, and is

Inserted into the posterior part of the fore-finger with the tendon of the common extensor, becoming closely united to it.

Situation: It arises nearer to the inner or ulnar edge of the arm than the extensor secundi internodii pollicis. It is concealed by the extensor digitorum communis, and extensor carpi ulnaris. The tendon passes under that of the common extensor.

Use: To assist in extending the fore-finger.

General Observations.—In dissecting the fore-arm, you must not expect to find each muscle separated from the contiguous ones, as far as its very origin from the bones. The partitions of fascia unite them most firmly to a considerable distance from their commencement: and these connexions should not be disturbed, since they are as justly to be regarded in the light of the origin of the muscles, as the attachments which the fibres have to the bone. The muscles are divided into two masses; of which one lies over the radius and back of the fore-arm, and contains the supinators and extensors: the other situated over the ulna and inner side of the fore-arm, consists of the pronators and flexors. Although some muscles, in each of these masses, arise quite separately, others are joined into a common tendinous and fleshy origin, in the way above described. First, clear all the muscles as they lie ; and after observing them in their relative position, reflect each muscle, beginning with the superficial ones, and reading its description at the same time.

SECTION III.

DISSECTION OF THE PALM OF THE HAND.

In the sole of the foot we saw a strong fascia, supporting and covering the muscles, and passing from the heel to the toes. On removing the integuments from the palm of the hand, we meet with a similar fascia, to which the integuments are firmly adherent. It arises from the tendon of the palmaris longus, and from the annular ligament of the wrist, expands over all the palm of the hand, and is fixed to the roots of the fingers, splitting to transmit their tendons. This is the FASCIA or APONEU-ROSIS PALMARIS. It is triangular. Where it arises from the wrist, it is narrow, and does not cover the bases of the metacarpal bones of the little and fore-finger.—As it descends over the hand, it becomes broader, and it is fixed by a bifurcated extremity to the lower end of each of the metacarpal bones of the four fingers.—This palmar fascia is strong and thick, conceals and supports the muscles of the hand, and may be distinguished into four portions, which are connected by transverse fibres; while other fibres pass downwards, adhere strongly to the edges of the metacarpal bones, and separate the tendons of each finger.

There is a small thin cutaneous muscle situated between the wrist and the little finger, on the inner edge of the hand.

The PALMARIS BREVIS—*Arises* from the annular ligament of the wrist, and from the inner edge of the fascia palmaris.

Inserted, by small scattered fibres, into the skin and fat which covers the short muscles of the little finger and inner edge of the hand.

Use: To make tense the palmar fascia, and to assist in contracting the palm of the hand.

The fascia palmaris may now be removed *. Under it will be seen the flexor tendons, coming from beneath the annular ligament of the wrist, and diverging as they pass towards their respective fingers. The tendons of the flexor sublimis are most superficial, and lie immediately over those of the flexor profundus. These flexor tendons, as they pass over the bones of the carpus into the palm of the hand, are firmly bound down by the

---

* Immediately beneath the palmar fascia is the superficial arterial arch, which is formed by the ulnar artery, and crosses over the tendons of the flexor sublimis; while the deep-seated arch, formed principally by the radial artery, lies close upon the metacarpal bones. The median and ulnar nerves also enter the palm of the hand under the palmar fascia;—for the description of these arteries and nerves, see the last section of the present chapter.

annular ligament; they lie enclosed, as in a channel; and a large bursa mucosa is found here, surrounding and sending processes between them :—this bursa extends from above the wrist to the metacarpus, lining the posterior surface of the annular ligament, and the front of the carpal bones. Having passed the annular ligament, the flexor tendons are surrounded by a loose cellular sheath, and are continued over the heads of the metacarpal bones, between the bifurcations of the palmar fascia, and are then lodged in the anterior concavity of the phalanges.

In their passage over the phalanges, the tendons are bound down by a fibrous sheath, and this sheath being laid open, it will be seen that each tendon of the flexor sublimis splits, at the extremity of the first phalanx, into two portions, for the passage of the tendon of the flexor profundus ; these portions re-unite, and again divide into two slips, to be attached to the base of the second phalanx ; and there is thus formed a kind of canal for the tendon of the profundus, which passes onwards to be inserted into the front of the base of the third phalanx.

The fibrous sheaths of the fingers should be examined; they will be found to consist of fibres, of a dehse shining appearance, which proceed transversely from the edges of the bones, binding down the tendons, and completing the *osseo-fibrous* canals, in which they glide. These have been termed *Vaginal Ligaments*, and they extend from the metacarpus to the middle of the extreme phalanx, being thick and strong over the middle part of the phalanges, but over the joints thin and indistinct ; the canal or passage is lined in its whole extent by synovial membrane, which is also reflected over the tendons themselves.

Connected with the tendons of the flexor profundus,

we find the Lumbricales, four small muscles, which arise, tendinous and fleshy, from the outer side of the tendons of the flexor profundus, soon after those tendons have passed the ligamentum carpi annulare. Each of these muscles has a small belly, which terminates in a tendon. The tendon runs along the outer or radial edge of the finger, and is

Inserted into the tendinous expansion which covers the back part of the phalanges of the fingers, about the middle of the first joint.

Use : To bend the first phalanges of the fingers, the flexor profundus being previously in action, to afford them a fixed point.

The Short Muscles of the Thumb and Fore-finger are five in number : they form the ball of the thumb and the fleshy mass between the thumb and fore-finger.

1. The Abductor Pollicis Manûs—*Arises*, by a broad tendinous and fleshy origin, from the anterior surface of the annular ligament of the wrist, and from the os scaphoides and os trapezium.

Inserted, tendinous, into the outer side of the root of the first phalanx of the thumb, and into the tendinous membrane which covers the back part of all the phalanges.

Situation : This is a flat, thin muscle, situated immediately under the integuments, and in the outermost portion of the muscular mass which forms the ball of the thumb.

Use : To draw the thumb from the fingers.

2. The Flexor Ossis Metacarpi Pollicis, or Opponens Pollicis—*Arises*, broad and fleshy, from the annular ligament of the wrist, and from the os scaphoides and os trapezium.

Inserted, tendinous and fleshy, into nearly the whole

4

length of the outer or radial border of the metacarpal bone of the thumb.

Situation : It lies under the abductor pollicis, and is almost entirely concealed : but a few of its fibres are seen projecting beyond the edge of that muscle.

Use : To bring the first bone of the thumb inwards.

3. The FLEXOR BREVIS POLLICIS MANÛS *arises* by two distinct heads.

(1.) The outer head *arises* from the inside of the annular ligament; from the anterior surface of the os trapezium and os trapezoides; and from the root of the metacarpal bone of the fore-finger.

Inserted into the outer sesamoid bone, and also into the base of the first phalanx of the thumb.

(2.) The inner head *arises* from the upper part of the os magnum and os unciforme, and from the root of the metacarpal bone of the middle finger.

Inserted, in like manner, into the inner sesamoid bone and base of the first phalanx of the thumb.

Situation : It is in great part concealed by the abductor pollicis : its inner origin is under the first lumbricalis : the two portions, of which this muscle consists, are separated above and below, but united behind the tendon of the flexor longus pollicis, for which they form a kind of channel. At their insertions, they are commonly united with the abductor and adductor muscles of the thumb.

Use : To bend the first joint of the thumb.

The tendon of the flexor longus pollicis, having passed between the two sesamoid bones, enters a fibrous sheath, and is invested by synovial membrane, (in the same manner as the flexor tendons of the fingers,) and proceeds to its insertion in the front part of the second phalanx.

4. The ADDUCTOR POLLICIS MANÛS—*Arises*, fleshy, from almost the whole length of the metacarpal bone sustaining the middle finger. The fibres converge, forming a triangular muscle, and pass over the metacarpal bone of the fore-finger, to be

Inserted, by a short tendon, into the inner part of the root of the first phalanx of the thumb.

Situation : The belly of this muscle is concealed ; it is deep-seated, and lies close to the bone under the tendons of the flexor profundus and lumbricales. The tendon is seen where it is inserted into the thumb, and runs along the inner edge of the flexor brevis pollicis.

Use : To draw the thumb towards the fingers.

5. The ADDUCTOR INDICIS MANÛS—*Arises*, tendinous and fleshy, from the os trapezium, and from the inner side of the metacarpal bone of the thumb. It forms a fleshy belly, runs over the side of the first joint of the fore-finger, and is

Inserted, by a short tendon, into the outer side of the root of the first phalanx of the fore-finger.

Situation : This muscle is seen most distinctly on the back of the hand, extending obliquely between the metacarpal bones of the thumb and fore-finger *. It is there superficial, and is crossed by the tendon of the extensor secundi internodii pollicis. In the palm of the hand it is concealed by the muscles of the ball of the thumb.

Use : To move the fore-finger towards the thumb, or the thumb towards the fore-finger.

---

* Many Anatomists consider the first of the Internal Interossei, or *Prior Indicis*, as a portion of this muscle, which is then described as arising by two heads, and as forming the first of the *external* or *bicipital interossei*. It is frequently termed *Abductor Indicis*, since it separates the fore-finger from the other fingers.

The insertion of the flexor carpi radialis is exposed by removing the muscles of the thumb: it is seen passing through its groove in the os trapezium to be inserted into the metacarpal bone of the fore-finger.

The SHORT MUSCLES of the LITTLE FINGER are three in number, and are situated on the metacarpal bone.

1. The ABDUCTOR MINIMI DIGITI MANÛS—*Arises* fleshy, from the os pisiforme, and adjacent part of the annular ligament of the wrist. Its fibres extend along the metacarpal bone of the little finger.

Inserted, tendinous, into the inner side of the first phalanx, and into the tendinous expansion which covers the back part of the little finger.

Situation: The belly of this muscle is superficial. It is only covered by the straggling fibres of the palmaris brevis, and is situated on the inner edge of the hand.

Use: To draw the little finger from the other fingers.

2. The FLEXOR PARVUS MINIMI DIGITI — *Arises,* chiefly tendinous, from the outer side of the os unciforme, and from the annular ligament of the wrist, where it is affixed to that bone: it descends, becoming narrower.

Inserted, by a roundish tendon, into the base of the first phalanx of the little finger.

Situation: This muscle is also covered by the fibres of the palmaris brevis. It lies on the inner side of the abductor minimi digiti, and its tendon is firmly connected to the tendon of that muscle. It is generally small and sometimes wanting.

Use: To bend the little finger, and bring it towards the other fingers.

3. ADDUCTOR METACARPI MINIMI DIGITI MANÛS, or OPPONENS MINIMI DIGITI—*Arises,* fleshy, from the os unciforme, and adjacent part of the annular ligament of the wrist. It forms a thick mass, which is

Inserted, tendinous, into the fore part of the meta-carpal bone of the little finger; nearly its whole length.

Situation: It is concealed by the bellies of the abductor and flexor brevis minimi digiti.

Use: To bend and bring the metacarpal bone of the little finger towards the palm.

The INTEROSSEI are small muscles situated between the metacarpal bones, and extending from the bones of the carpus to the fingers. They are exposed by removing the other muscles of the thumb and fingers.

The INTEROSSEI INTERNI are seen in the palm of the hand, and are four in number. They *arise*, tendinous and fleshy, from the base and sides of the metacarpal bones, and are inserted into the sides of the first phalanx of the fingers, and into the tendinous expansion which covers the posterior surface of all the phalanges.

1. The First, named Prior Indicis, *arises* from the outer part of the metacarpal bone of the fore-finger; and is *inserted* into the outer side of the first phalanx of that finger; *Use*: To draw the fore-finger towards the thumb. This muscle, lying along the radial edge of the fore-finger, belongs as much to the back as to the front of the hand; it is connected with the adductor indicis, and is often regarded as one of its heads.

2. The Second, named Posterior Indicis, *arises* from the root and inner side of the metacarpal bone of the fore-finger; and is *inserted* into the inner side of the first phalanx of the fore-finger. *Use*: To draw that finger outwards.

3. The Third, named Prior Annularis, *arises* from the root and outer side of the metacarpal bone of the ring finger; and is *inserted* into the outer side of the first phalanx of the same finger. *Use*: To pull the ring-finger towards the thumb.

4. The Fourth, named Interosseus Auricularis, *arises*, from the root and outer side of the metacarpal bone of the little finger: and is *inserted* into the outer side of the first phalanx of the little finger. *Use:* To draw the little finger outwards.

The internal interossei also assist in extending the fingers obliquely.

The INTEROSSEI EXTERNI, or BICIPITES, are three in number. They are larger than the internal, and are situated betwixt the metacarpal bones on the back of the hand. Each of these muscles *arises* by a double head, from two metacarpal bones, and is inserted into the side of the first phalanx of one of the fingers, and into the tendinous expansion which covers the posterior part of the phalanges.

1. The First, named Prior Medii, *arises* from the roots of the metacarpal bones of the fore and middle fingers; and is *inserted* into the outer side of the middle finger. *Use:* To draw the middle finger towards the thumb.

2. The Second, named Posterior Medii, *arises* from the roots of the metacarpal bones of the middle and ring-fingers, and is *inserted* into the inner side of the middle finger. *Use:* To draw the middle finger towards the ring-finger.

3. The Third, named Posterior Annularis, *arises* from the roots of the metacarpal bones of the ring and little fingers; and is *inserted* into the inner side of the ring-finger. *Use:* To draw the ring-finger inwards. The external interossei also extend the fingers.

The back of the hand and extensor tendons are covered by a very thin and transparent expansion, which is apparently continued from the fascia of the fore-arm.

SECTION IV.

OF THE VESSELS AND NERVES OF THE SUPÉRIOR EXTREMITY.

I. ARTERIES.

THE Subclavian and Axillary divisions of the great arterial trunk of the upper extremity have been described. When it has passsd over the edge of the latissimus dorsi and teres major, it takes the name of Brachial Artery.

The BRACHIAL ARTERY may be said to have its course along the inside of the arm, crossing the shaft of the humerus obliquely, to reach the inner and fore part of the fold of the elbow. Having left the axilla, it runs along the inferior edge of the coraco-brachialis, being separated from the triceps behind by some cellular tissue and the spiral nerve. Rather higher up than the middle of the os humeri, it crosses over the tendinous insertion of the coraco-brachialis, lying close upon the bone, and being here situated between the belly of the biceps flexor cubiti, and the superior fibres of the brachialis externus. The artery then passes along and rather behind the inner edge of the biceps flexor, descending betwixt that muscle and the fibres of the brachialis internus. In dissecting this vessel, we find it invested by a fascia or sheath, which is little more than condensed cellular membrane. This fascia may be traced extending from the internal intermuscular ligament. It covers the brachial artery and median nerve, and the great basilic vein, as it enters the axilla, is found lying in the fore and inner part of the sheath. On dissecting this fascia, we find, close to the margin of the coraco-brachialis and biceps flexor, the median nerve; under it the brachial artery, with its two venæ comites, one on each side; and more superficially seated, the basilic vein. As the artery approaches the lower extremity of the os humeri, it inclines forwards toward the fold of the arm, and dives beneath the aponeurosis, which arises from the inside of the tendon of the biceps flexor. Its situation at the fold of the arm has been described.

In this course down the arm, the brachial artery is covered only by the integuments and fascia of the arm, lying deeper and internally above, but, as it descends, becoming more superficial and anterior. The Median nerve crosses obliquely over the artery, and then runs along its inner or ulnar side to the fold of the elbow; the Ulnar nerve is posterior to the artery, and separated from it, in the inferior third of the arm, by the intermuscular ligament; while the Internal Cutaneous nerve descends superficially in the line of the artery.

BRANCHES OF THE BRACHIAL ARTERY.

(1.) *Arteria Profunda Humeri Superior* is sent off from the inner side of the brachial artery, immediately where it has left the fold of the arm-pit. It passes downwards and backwards round the os humeri, and is accompanied by the musculo-spiral or radial nerve. It takes its course beneath the triceps extensor, and will be found descending betwixt the two shorter portions of that muscle, which have their origin from the humerus. It gives off some muscular twigs, and divides on the posterior surface of the humerus into two branches. One descends beneath the triceps extensor towards the olecranon, and communicates with the ulnar and interosseous recurrents; the other, which is the continued trunk, winds with the nerve round the humerus, coming out at the interval between the triceps and brachialis internus, and descends towards the outer condyle, communicating with the radial recurrent.

(2.) The *A. Profunda Inferior* is smaller than the last, and is sent off from the brachial artery about two inches lower down. It descends along the inside of the arm, in the direction of the ulnar nerve, piercing the internal intermuscular ligament; it terminates in ramifying about the inner condyle, communicating with the ulnar recurrents; one branch descends, with the ulnar nerve, in the fossa between the inner condyle and olecranon.

(3.) The *Ramus Anastomoticus Major* passes off from the inner side of the brachial artery, about two or three inches above the inner condyle; it is very tortuous, and passes inwards on the brachialis internus; it gives off muscular twigs, and some which pass over the fore part of the inner condyle; but the chief branch perforates the internal intermuscular ligament, and communicates, in the space between the inner condyle and olecranon, with the posterior ulnar recurrent, and with the descending branches of the arteriæ profundæ.

(4.) *Muscular branches* come off from the brachial artery in its whole course down the arm. About the middle of the arm, it gives off the *arteria Nutritia* of the humerus, which sends off some muscular twigs before it enters the bone.

The Brachial Artery, where it lies deep under the aponeurosis of the biceps, divides into its two branches. 1. The Radial, and 2. The Ulnar, which last gives off a third principal branch, the Interosseous artery.

1. The RADIAL ARTERY is smaller than the ulnar, and in its course more superficial. It leaves the ulnar artery, and inclines towards the radial or outer edge of the forearm, descending nearly in a line from the middle of the fold of the arm to the front of the styloid process of the radius. At first it lies betwixt the pronator teres, and supinator longus, and is covered anteriorly by the latter

muscle. It then descends close along the inner edge of the supinator longus, and about the middle of the fore-arm passes over the insertion of the pronator teres. It then holds its course, first on the fibres of the flexor digitorum sublimis, next on the flexor longus pollicis and pronator quadratus, having still the supinator longus on its outer side, and on its inner side the flexor carpi radialis. It is here more superficial, and is felt pulsating beneath the integuments, between the tendons of the two last-named muscles, for some considerable distance above the wrist. Having arrived at the lower extremity of the radius, it turns over the outer side of the carpus to the back of the hand. It passes over the external lateral ligament of the wrist, beneath the chord formed by the two first extensors of the thumb, is again felt beating in the fossa between these two tendons and the third or extensor major, then passing forwards over the junction of the os trapezium and trapezoides, and, arriving at the space betwixt the bases of the metacarpal bones of the thumb and fore-finger, it plunges into the palm of the hand.

In this course, the Radial artery is usually accompanied by two veins, and, in the upper part of the fore-arm, it has on its outer side the anterior branch of the musculo-spiral or radial nerve. It is not immediately covered by any muscle, but by the general fascia of the fore-arm, and by a deeper-seated layer of fascia, which may be seen binding it down to the muscles over which it crosses.

The branches of the radial artery, in its course along the fore-arm, are the following:

(1.) A. *Radialis Recurrens* is sent off from the radial immediately after it leaves the ulnar artery, and is of considerable size; it passes at first transversely outwards towards the supinator longus, giving many branches to the adjacent muscles,—then bends upwards in front of the external condyle of the humerus, and terminates in inosculating with the profunda superior.

(2.) *Muscular branches*, small, but numerous, are given off by the radial artery in its course down the arm : one of these, small and deep-seated but constant, is worthy of remark : it springs from the artery near the wrist, and passes inwardly to the pronator quadratus.

(3.) A. *Superficialis Volæ*, a branch of variable size, comes off from the radial artery, just before it passes to the back of the hand, sometimes higher up : it runs superficially downwards in front of the anterior annular ligament to the palm of the hand, generally perforating the upper extremity of the abductor pollicis. It is distributed to the muscles of the thumb and integuments, and anastomoses 'with the extremity of the superficial Palmar arch of the ulnar artery.

Having turned to the back of the hand, the Radial artery gives off

(4.) Some Dorsal branches. (*a.*) A. *Dorsalis Carpi*, which crosses obliquely the carpus. (*b.*) A. *Dorsalis Metacarpi*, over the second metacarpal bone and back of the hand. (*c.*) A. *Dorsales Pollicis*, from one to three branches, to the back of the metacarpal bone and phalanges of the thumb, and a branch from which is generally sent to the index finger.

Having reached the palm of the hand, the radial artery divides into two large branches, viz.

(5.) A. *Magna Pollicis*, which sends two arteries along the anterior part of the thumb, and also generally gives off the A. *Radialis Indicis*, which passes along the outer edge of the fore-finger, and inosculates with a branch of the ulnar artery.

(6.) The trunk of the radial artery forms the *Deep-seated Palmar Arch*. This arch passes from the root of the thumb across the metacarpal bones near their bases, and terminates at the metacarpal bone of the little finger, inosculating with the communicating or deep palmar branch of the ulnar artery. This arch lies deep, close to the bones, under the flexor tendons. It supplies the interosseous muscles and deep-seated parts of the palm, and some of its branches pass betwixt the metacarpal bones to the back of the hand.

2. The ULNAR, or CUBITAL ARTERY, is the largest of the two branches of the Brachial. It takes its course deep among the muscles on the inside of the fore-arm. Its direction is somewhat curved, first inclining inwards from the fold of the arm to the inner border of the ulna above its middle, then descending in a direct line to the os pisiforme. It is seen passing under the pronator teres, flexor carpi radialis, palmaris longus, and flexor sublimis perforatus, but over the flexor profundus perforans. It descends in the connecting cellular membrane, between the flexor sublimis and profundus; but, about the middle of the fore-arm, it emerges from these muscles, and appears at the ulnar edge of the arm, betwixt the flexor sublimis and flexor carpi ulnaris, resting on the fibres of the flexor profundus. It is here superficially seated, being only covered by the integuments and general fascia of the fore-arm, and by a second layer of fascia, which passes from the tendon of the flexor ulnaris to the flexor profundus. It passes into the palm of the hand over the annular ligament of the wrist, lying external to the ligament, and on the outer or radial side of the os pisiforme, but it is covered by the slip of fascia which ties down the tendon of the flexor carpi ulnaris. It then sinks under the palmar aponeurosis, and arriving at the base of the metacarpal bone of the little finger, bends outwards to form the Superficial Palmar Arch.

In this course the artery has two veins accompanying it, one on each side, and in the lower two-thirds of the fore-arm, the ulnar nerve lies close to the inner side of the artery, and where it passes over the wrist, the nerve is placed between the artery and the os pisiforme.

The SUPERFICIAL PALMAR ARCH lies over the flexor tendons, immediately beneath the palmar aponeurosis. It crosses the metacarpal bones obliquely, betwixt their bases and the middle of their bodies, beginning at the root of the little finger, and terminating, at the root of the thumb, in inosculations with the branches of the radial artery. The convex side of the arch is turned towards the fingers, and sends off five branches.

(1.) A *branch* to the muscles and inner edge of the little finger.

(2.) *Ramus Digitalis Primus*, or the first digital ar-tery, which runs along the space betwixt the two last metacarpal bones, and, arriving at the roots of the pha-langes, bifurcates into two branches, one to the outside of the little finger, and the other to the inner side of the ring-finger.

(3.) The *Second Digital Artery*, which bifurcates in a similar manner, and supplies the outer edge of the ring-finger, and the inner side of the middle finger.

(4.) The *Third Digital Artery*, which is distributed to the outer edge of the middle finger, and to the inner side of the fore-finger.

(5.) The *last branch* may be regarded as the termi-nation of the superficial arch, anastomosing near the ball of the thumb with the A. superficialis volæ and radialis indicis. The superficial arch sometimes supplies the thumb and fore-finger with the branches usually given off from the radial artery.

. From the concavity of the arch are sent off the *Inter-*

4

osseous Arteries of the palm, small twigs which supply the deep-seated parts, and perforate betwixt the metacarpal bones to the back of the hand.

The branches of the ulnar artery, in its course along the fore-arm and wrist, are the following.

(1, 2.) The *Recurrent Arteries* are two in number. They are sent off from the ulnar artery immediately below the elbow, sometimes in one common branch, which subdivides. The *Anterior Recurrent* is small, and ascends towards the fore part of the inner condyle. The *Posterior Recurrent* is a much larger artery,—it passes upwards and backwards, behind the pronator teres, palmaris and flexor sublimis, and before the flexor profundus; it then ascends behind the inner condyle, in the hollow between it and the olecranon, by the side of the ulnar nerve, passing between the two origins of the flexor ulnaris, and it there terminates in a free anastomosis with the inferior and superior Profundæ and Ramus anastomoticus.

(3.) The *Interosseous Artery* is the next branch.

(4.) Numerous *muscular branches* pass off from the ulnar artery in its course down the arm; they are small and irregular; one branch longer than the rest, descends behind the median nerve to the lower part of the fore-arm.

(5.) *A. Dorsalis Carpi* is sent off from the ulnar artery, a little above the wrist to the back of the hand. Twigs are also given off to the annular ligament and neighbouring parts.

(6.) *A. Palmaris Profunda,* or communicating branch, is sent off from the ulnar artery, where it descends by the side of the os pisiforme. It sinks into the flesh at the root of the little finger, and inosculates with the termination of the deep-seated palmar arch of the radial artery.

3. The Interosseous Artery. This third principal artery of the fore-arm comes off from the ulnar, immediately after the ulnar recurrent. It passes backwards, towards the interosseous space, and shortly divides into two branches.

(1.) The *External*, or *Posterior Interosseous Artery*, is the smallest branch. It passes through the upper part of the interosseous ligament, and thus reaches the back part of the arm, under the anconeus muscle: here it bifurcates into two nearly equal branches, of which one, the *Interosseous Recurrent*, ascends between the anconeus and extensor ulnaris, and between the external condyle and olecranon, to the triceps extensor:—The other, or proper *Posterior Interosseous artery* descends between the two layers of muscles on the back part of the fore-arm to the carpus.

(2.) The *Anterior Interosseous Artery* runs down close upon the middle of the interosseous ligament, betwixt the flexor longus pollicis and flexor profundus perforans, giving numerous branches to the adjacent muscles. Arriving at the upper edge of the pronator quadratus, it perforates betwixt the radius and ulna to the back part of the arm, and spreads its extreme branches on the wrist and back of the hand.

The Brachial Artery and its branches are subject to some variations: the principal are the following.

1. The Brachial artery may divide high up in the arm, sometimes as high as the axilla. The smaller branch is generally the Radial Artery, and it takes its course superficially over the elbow-joint.

2. The Radial Artery frequently turns over the radius at some distance above its lower extremity.

3. The A. Superficialis Volæ is not unfrequently a large artery, and then commonly joins the ulnar artery in forming the superficial palmar arch.

II. VEINS.

The Cutaneous Veins have been already described.

The Brachial Artery is accompanied by two veins, named Venæ Comites, or Satellites. These receive branches corresponding to the ramifications of the artery, and are continued into the axillary vein ; but, in the palm of the hand, the veins are not satellites of the arteries, the Digital veins chiefly forming the roots of the superficial or cutaneous veins.

III. NERVES.

In the dissection of the axilla, we demonstrated the great axillary plexus, and traced its first branches ramifying about the shoulder and chest. The distribution of the five remaining branches of the plexus must now be described.

1. The INTERNAL CUTANEOUS NERVE is chiefly derived from the first dorsal and last cervical nerves. It has already been noticed among the cutaneous nerves of the arm, coming out from the axilla, and descending along the inner side of the arm with the basilic vein, in the direction of the brachial artery. It is the smallest of the nerves of the arm, and, after detaching some inconsiderable filaments, divides above the elbow-joint, sometimes higher up, into two branches. The *External* follows the border of the biceps flexor, and crosses over the middle of the fold of the arm, to ramify above the fascia. The *Internal*, which seems the trunk of the nerve, accompanies the basilic vein, and divides, near the inner condyle, into filaments, which descend over the front and inner side of the elbow-joint, to the integuments on the fore, inner, and back part of the fore-arm.

2. The EXTERNAL CUTANEOUS NERVE, Musculo-Cutaneus, or Perforans Casserii, is rather larger than the last, and comes principally from the fifth and sixth cer-

vical nerves. It passes downward, generally piercing through the fibres of the coraco-brachialis muscle. After this passage, it continues its course obliquely across the arm, betwixt the biceps flexor and the brachialis internus. It gives twigs to both these muscles, and emerges from beneath the outer edge of the biceps, to appear as a superficial nerve on the front edge of the supinator longus. It then passes over the fold of the elbow, under the median cephalic vein, and descends along the outer or radial side of the fore-arm, between the fascia and skin, giving off numerous cutaneous twigs, as far as the root of the thumb and back of the hand.

3. The MEDIAN NERVE, (sometimes called the Radial,) is the largest of the branches of the axillary plexus, and is formed by fasciculi from all the nerves composing the plexus. It accompanies the brachial artery to the bend of the elbow; it is felt as a firm chord in the line of the artery, and is contained in the same cellular sheath or canal. In its passage down the arm, it lies at first before the artery, or somewhat to its outer side; but, in its progress downwards, it crosses over it, and at the elbow is situated on its inside. It gives off no branches until it has sunk, with the artery, under the aponeurotic expansion of the biceps flexor. Here it distributes many nerves to the muscles of the fore-arm, to the pronator teres, flexor carpi radialis, the flexors of the thumb and fingers, and a considerable filament, the *Nervus Interosseus*, which descends with the anterior interosseous artery, upon the interosseous ligament, and, detaching a few filaments to the pronator quadratus, perforates with the artery, above the upper edge of that muscle, to the back of the carpus. The trunk of the Median Nerve perforates the pronator teres, passes betwixt the flexor digitorum sublimis and flexor profundus, and descends, in the middle of the arm, betwixt these

muscles down to the wrist. Near the wrist it becomes more superficial, lying amongst the tendons of the flexors, and before it descends under the annular ligament, sends a superficial, or *Cutaneous Palmar* branch, to the integuments and short muscles of the thumb. The nerve itself passes with the flexor tendons of the fingers under the annular ligament of the wrist, emerges from these tendons, and appears on their outside, near the root of the thumb. It ramifies superficially in the hand, setting off five branches *.

(a) The first branch passes to the short muscles and outer or radial edge of the thumb.

(b) The second branch proceeds to the inner or ulnar border of the thumb.

(c) The third to the side of the fore-finger, next the thumb.

(d) The fourth branch descending between the second and third metacarpal bones, subdivides into two nerves, of which one passes to the inner side of the fore-finger, and the other to the outer side of the middle finger.

(e) The fifth also subdivides into two, to the inside of the middle finger, and outer side of the ring-finger.

These nerves proceed forwards under the palmar fascia; they pass before the flexor tendons, but behind the superficial palmar arterial arch, to reach the fingers, and accompany the digital arteries.

4. The ULNAR, or CUBITAL NERVE, is chiefly formed from the two last branches of the axillary plexus, and comes off from its lower or back part. It descends on the inside of the arm, along the inner border of the triceps extensor muscle. It is at first situated imme-

---

* This distribution sometimes takes place by two principal branches, which subdivide, the external into three, the internal into two branches.

diately under the integuments and brachial fascia, but below the middle of the arm it is tied down by the intermuscular ligament, which passes to the inner condyle of the humerus, above which it gives off some slender cutaneous filaments. The nerve here becomes more deeply seated ; it runs behind the inner condyle in the hollow between it and the olecranon, and in the flesh of the brachialis externus, or third head of the triceps extensor. After passing the condyle, it continues its course betwixt the two heads of the flexor carpi ulnaris, till it reaches the ulnar artery. It then accompanies the ulnar artery, lying on its inside, and running along the fore-arm betwixt the flexor ulnaris and flexor digitorum profundus. It sends twigs to the neighbouring muscles, and when arrived near the wrist, divides into two branches.

(1.) The smaller branch, called *Posterior*, or **Dorsal**, passes under the tendon of the flexor carpi ulnaris, and over the lower end of the ulna, to be distributed to the back of the hand, and of the little and ring-fingers.

(2.) The continued trunk of the nerve accompanies the ulnar artery over the annular ligament of the wrist, lying close to the os pisiforme, and, before reaching the palm of the hand, divides into two branches, one deep, the other superficial.

(*a.*) The *deep branch* sinks under the flexor tendons, and bends outwards, forming a kind of deep-seated arch, which gives filaments to the interosseous muscles, and terminates in the short muscles of the thumb and fore-finger.

(*b.*) The *superficial branch* gives filaments to the muscles of the little finger, and subdivides into a branch which passes to the inner side of the little finger, and a larger branch, which passes under the palmar aponeurosis, and is distributed to the outer side of the little finger and inner side of the ring-finger, communicating with the median nerve.

5. The MUSCULO-SPIRAL, or RADIAL NERVE is formed in general by fasciculi of the three last cervical nerves and first dorsal. It equals or exceeds in size the median nerve, and passes from the axilla behind the os humeri, making a spiral turn round the bone to reach the outside of the arm. It first descends in the interval between the two humeral portions of the triceps extensor, accompanying the arteria profunda superior, and passing deep into the flesh of the arm. It gives many branches to the muscles of the arm, particularly the triceps, and, while still behind the bone, a remarkable *cutaneous nerve* *, which pierces in general the outer fibres of the triceps, comes out upon the supinator longus, and descends, superficially, upon the radial edge of the fore-arm. From the back part of the arm, the great trunk of the spiral nerve is reflected spirally forwards. It is found emerging betwixt the supinator longus and brachialis internus, seated deep and close to the bone. It descends betwixt these muscles, keeping close to the edge of the supinator longus, and passing over the elbow-joint; and at this point, while it is still lying deep, between the brachialis internus and radial extensors, the trunk of the spiral nerve divides itself into two branches.

(1.) The first, *Posterior*, or **Deep** branch is the larger; it perforates the fleshy fibres of the supinator brevis, giving many filaments to this and the neighbouring muscles, and, reaching the back part of the fore-arm, supplies the extensor muscles of the hand and fingers: one considerable filament, (*N. interosseus externus*) descends on the interosseous membrane, under the annular ligament, to the back part of the hand.

* *Nervus cutaneus externus superior.*

(2.) The *Anterior*, or *Superficial* branch, or proper *Radial Nerve*, accompanies the supinator longus down the fore-arm, lying within its anterior border, and on the outer side of the radial artery. At the lower third of the fore-arm, it turns outwards over the edge of the radius, passing beneath the tendon of the supinator longus, and then descends between the integuments and the tendons of the two first extensors of the thumb. It soon divides into two branches, which are distributed to the back of the wrist, thumb, both sides of the fore-finger, and external side of the middle-finger.

CHAPTER XVI.

DISSECTION OF THE JOINTS.

THE examination of the Joints in the recent state usually follows that of the muscles, and requires much patient and careful proceeding. It may be remarked generally, that all the soft parts should be removed, except the ligaments which surround the articulations. Caution, however, is required, as some of the neighbouring tendons send expansions over the joints, which intermix with the proper ligaments; and this connexion, as well as the manner in which the articulations receive support from the muscles passing over them, should be observed. The Ligaments themselves are distinguished by their white colour, and dense and firm texture.

SECTION I.

OF THE ARTICULATIONS OF THE TRUNK AND HEAD.

§. 1. ARTICULATIONS OF THE VERTEBRÆ IN GENERAL.

EACH Vertebra, from the third cervical inclusive to the last lumbar, is articulated with the one preceding, and that following it, by its body and articulating processes. The arches and spinous processes of the vertebræ are also connected by ligaments.

Cut out, with the saw, a portion of the vertebral column; this is the most convenient mode of examination.

1. The BODIES of the Vertebræ are united by two ligaments, and by the intervertebral substance:

(1.) *Common Anterior Vertebral Ligament*, or Fascia Longitudinalis Anterior, is exposed by simply denuding the front of the spinal column. It extends from the second cervical vertebra to the first bone of the sacrum, adhering to the fore part of the bodies of the vertebræ and to the intervertebral substances; it is narrow in the neck, broad in the back, and still broader in the loins, where it is strengthened by the tendinous fibres of the crura of the diaphragm.

(2.) The *Common Posterior Vertebral Ligament*, or Fascia Longitudinalis Postica, extends, in like manner, along the posterior surface of the bodies of the vertebræ, within the vertebral canal, which must be cut open to expose it. It is of a shining pearly-white appearance, narrow over the bodies of the vertebræ, but expanded on the intervertebral cartilages, descending from the second cervical vertebra to the sacrum, and being, posteriorly, in contact with the dura mater investing the spinal marrow.

(3.) The *Inter-vertebral substances* are placed between the bodies of the vertebræ, being closely adherent to the bony surfaces; of an elastic fibro-cartilaginous structure, lamellated, with a soft gelatinous substance interposed, particularly in the centre of each intervertebral substance, which is quite pulpy. Some cross slips of ligament, passing between the vertebræ over the intervertebral substances, have been named *Crucial Ligaments.*

2. The ARTICULATING PROCESSES of the Vertebræ are

united by small *Synovial Capsules*, which are strengthened externally by irregular ligamentous fibres.

3. The ARCHES or RINGS of the vertebræ are not in
contact, but are united by the *Ligamenta Subflava*:
these are most apparent within the vertebral canal; they
are formed of dense, elastic fibres, of a yellow colour,
filling up the space between the rings from the second
vertebra to the sacrum, and thus completing posteriorly
the spinal canal.

4. The SPINOUS PROCESSES of the dorsal and lumbar
vertebræ are connected by the *Interspinous Ligaments*,
which in the neck are wanting:—they have been divided
into *membrana interspinalis*, a thin expansion between
the bodies of the spinous processes, and *funiculi liga-
mentosi* extending between the apices of these processes.

The transverse processes of the dorsal vertebræ from
the fifth to the eleventh are connected by small irregular bundles of ligamentous fibres.

The *Superficial Cervical Ligament*, or ligamentum
nuchæ, has been described in the dissection of the neck,
and is chiefly of use to give attachment to muscles.

§. 2. ARTICULATION OF THE OCCIPUT WITH THE FIRST
AND SECOND VERTEBRÆ, AND OF THE VERTEBRÆ WITH
EACH OTHER.

Remove the os occipitis with the three or four first
vertebræ attached to it:—then expose the vertebral canal by dividing vertically the bony arches of the vertebræ, and carry on the saw in the same direction, so as
to cut through the occipital bone and lay open the foramen magnum from behind.

1. The CONDYLES of the Os OCCIPITIS articulate with
the superior articular cavities of the ATLAS; the connecting ligaments between the two bones are

(1.) The *Anterior Ligament*, a broad expansion,

which extends from the anterior arch of the atlas to the anterior edge of the foramen magnum, and which is strengthened, in the median line, by a distinct prominent fasciculus, descending from the basilar process to the anterior tubercle of the atlas.

(2.) The *Posterior Ligament*, a similar broad expansion, continued from the posterior arch of the atlas to the back part of the foramen magnum : it consists of two laminæ, of which the anterior is interlaced with the dura mater.

(3.) *Synovial Capsules* of the articulations of the condyles.

2. The Os occipitis is connected with the DENTATA or second vertebra, intermediately, by ligaments, which are situated within the vertebral canal.

(1.) On removing the dura mater, a broad flat ligament is seen, passing from the basilar process of the os occipitis, behind the odontoid process, to be inserted into the body of the second and third vertebræ, intermixing with the posterior vertebral ligament, of which some consider it to be a portion : this is the *Apparatus Ligamentosus* of Winslow.

(2.) Remove carefully the preceding ligament ; you expose the *Odontoid* or *Lateral Ligaments ;* two short, thick, strong ligaments, of a rounded form, which extend obliquely from the sides and point of the odontoid process to the edge of the foramen magnum, and to the rough fossa at the inner side of the occipital condyles. Some slips of fibres between these two lateral ligaments have been termed the *Perpendicular Ligament.*

3. The ATLAS is connected with the DENTATA : 1st, By the odontoid process of the dentata articulating with the anterior arch of the atlas, and with its transverse ligament. 2d, By the lateral articulating surfaces of the two vertebræ,

(1.) The ODONTOID PROCESS has two smooth surfaces, corresponding, in front, to the back part of the
anterior arch of the atlas, and, posteriorly, to the transverse ligament. The ligaments are;

(a.) The *Transverse Ligament*, a thick flattened chord,
extending across the atlas, from one side to the other,
behind the odontoid process, and forming with the anterior arch of the atlas a sort of ring, in which the
odontoid process turns. This ligament has two small
vertical slips or *appendices;* the superior extending upwards to the foramen magnum,—the inferior fixed to
the root of the odontoid process.

(b.) A *Synovial Capsule*, loose and transparent, embraces each of the articular surfaces just described of
the odontoid process.

(2.) The lateral ARTICULATING PROCESSES of the two
Vertebræ, have *Synovial Capsules*, which are remarkable for their laxity: the two bones are maintained in
relation by an *Anterior ligament*, which descends from
the inferior border of the atlas and from its anterior
tubercle, to the base of the odontoid process and front of
the dentata; and by a *Posterior ligament*, membranous,
and very loose, extending between the posterior arches
of the two bones.

There is no intervertebral substance between the atlas
and dentata.

§. 3. ARTICULATIONS OF THE RIBS.

The RIBS are articulated with the VERTEBRÆ: 1st,
By their HEADS, which have articular surfaces, and are
received into the cavities, formed by two adjoining vertebræ and intervertebral substance; and, 2nd, by their
TUBERCLES, which are covered with a thin cartilage, and
articulate with the summit of the TRANSVERSE PRO-
CESSES. The first, eleventh, and twelfth ribs are receiv-

ed into cavities formed by single vertebræ, and the tubercles of the two last ribs are not articulated.

Remove some of the middle dorsal vertebræ, with the posterior extremities of the ribs attached; clear away the soft parts not connected with the articulations.

The Ligaments of the COSTO-VERTEBRAL articulation are,

(1.) The *Anterior* or *Radiated Ligament*, which is seen within the chest, immediately beneath the pleura, extending from the head of the rib, over the articulation, to be inserted by three separate fasciculi; the superior, into the body of the vertebra above; the inferior into the vertebra below, and the middle fasciculus passing transversely to the intervertebral substance, (Ligamenta capitelli costæ.)

(2.) The *Inter-articular Ligament* is exposed on cutting open the articulation,—a fibrous fasciculus, passing from the intervertebral substance to the head of the rib, and dividing the articulation into two cavities, which have

(3.) Distinct *Synovial Capsules.*

The Ligaments of the COSTO-TRANSVERSE articulation are,

(1.) The *External* or *Posterior Transverse Ligament*, which is seen, on the back part of the rib, passing nearly transversely from the apex of the transverse process of the vertebra, to the outer side of the tubercle of the rib.

(2.) The *Ligamentum Cervicis Costæ*, or Ligament of the neck of the rib, is a strong ligamentous band, passing from the inferior part of each transverse process; to the upper part of the neck of the rib, below that with which it is articulated. It consists generally of two fasciculi, the one *internal*, the other *external*, which cross each other. The first and last ribs are destitute of this ligament.

(3.) A small *Synovial Capsule.*

The CARTILAGES of the seven true Ribs are articulated with the sternum. They are received into cavities in the lateral edge of the sternum, a very delicate synovial membrane covers the articular surfaces, and ligamentous fibres proceed in front and behind.

(1.) The *Anterior Ligaments*, broad and triangular, interlace with each other, and, assisted by the aponeurotic fibres of the pectoralis major, form an expansion, which covers the whole anterior surface of the sternum.

(2.) The *Posterior Ligaments* form in the same manner, by the interlacing of their fibres, a *distinct membrane* investing the posterior surface of the sternum.

. (3.) The Cartilage of the seventh rib is united to the ensiform cartilage by a particular ligament, the *Costo-xiphoid*.

The Cartilage of the first rib is inseparably united with the first bone of the sternum. The cartilages of the sixth, seventh, eighth, and ninth ribs are connected mutually by loose synovial capsules and irregular ligamentous fibres. Between the two next ribs we only find some tendinous fibres, and the last rib is connected solely with the abdominal muscles.

The three pieces, of which the STERNUM is composed in the adult, are respectively connected by a plate of fibro-cartilage, placed between their contiguous borders: this sometimes disappears between the second bone and ensiform cartilage, but between the two upper bones it is rarely obliterated, except in advanced life. The union is supported by tendinous fasciculi, which intermix with the anterior and posterior sternal membranes.

§. 4. ARTICULATIONS OF THE PELVIS.

To examine the Ligaments of the Pelvis, little other preparation is required than merely dissecting off the muscles.

1. Sacro-Vertebral Articulation. The Sacrum articulates with the fifth or last lumbar vertebra, exactly in the same manner as the vertebræ are united mutually; in addition there are two peculiar ligaments,

- (1.) The *Sacro-Vertebral Ligament*, strong and short, arising from the lower and anterior part of the transverse process of the last lumbar vertebra, and descending obliquely outwards to be fixed into the base or upper part of the sacrum: it is covered anteriorly by the psoas muscle.

(2.) The *Ilio-Lumbar Ligament* connects mediately the ilium with the last lumbar vertebra. It is of a flat triangular shape, covered posteriorly by the sacro-lumbalis and longissimus dorsi muscles,—anteriorly, by the psoas magnus. It is fixed to the inner part of the posterior tuberosity of the ilium, and passes transversely outwards to the transverse process of the fifth and sometimes the fourth lumbar vertebra.

2. Sacro-Iliac Articulation or Symphysis. This is formed by the union of corresponding surfaces of the sacrum and ilium : there is an intervening *fibro-cartilage*, adhering to both bones, which is seen on tearing them asunder ; and the articulation is further strengthened by ligaments—

(1, 2.) The *Greater* and *Less Sacro-Sciatic Ligaments*, which belong to the outlet of the pelvis, rather than to the articulation, and have been described in the dissection of the thigh.

(3.) There is a distinct, and strong, flattened ligament, placed nearly vertically behind the articulation, covered by the gluteus maximus, and connected with the great sacro-sciatic ligament. This has been called the *Sacrospinous Ligament* ; it is fixed, by one extremity, to the upper and back part of the spine of the ilium, and, by

s

the other, to the lateral and posterior parts of the sacrum.

(4.) The *Sacro-iliac Ligament* consists of irregular but strong ligamentous fibres, passing from the upper back part of the sacrum to the inner surface of the tuberosity of the ilium, occupying the interval between the two bones behind, under the mass of lumbar muscles...

Besides these, there are many ligamentous bands, which assist in connecting the sacrum and ilium : those in front of the articulation are irregularly disposed, and are of a bright shining appearance.

3, Sacro-Coccygean Articulation. The lower end of the sacrum is united to the base of the coccyx by a thin intervertebral substance, and by anterior and posterior ligamentous bands ; of these the posterior are more distinct, passing over and closing the termination of the vertebral canal.

4. The Symphysis Pubis is formed by the union of the two oval surfaces of the ossa pubis, with a fibro-cartilaginous plate interposed, which is much thicker anteriorly than behind. Ligamentous fibres pass across strengthening the articulation : (1.) On the *anterior* surface, this ligamentous covering is interlaced with the aponeuroses of the abdominal muscles and with the periosteum. (2.) The *Posterior ligament* extends itself into the arch of the pubis, for half an inch below the symphysis; where it is very strong and thick, and has been named the *Pubic Ligament.*

5. The Obturator Ligament is a fibrous membrane, filling up the circumference of the obturator foramen, except at its upper part, where the notch exists, and where it leaves an opening for the passage of the obturator vessels and nerve.

§. 5. Articulation of the Lower Jaw.

This is formed, on each side, by the Condyle of the Inferior Maxilla being articulated with the Glenoid Cavity of the temporal bone and with the articular eminence before it.

Remove the soft parts, which surround the articulation.

The Ligaments are,

(1.) The *External Lateral Ligament*, a flattened, very short, fibrous band, passing from the tubercle at the root of the zygomatic process to be fixed, inferiorly, to the external side of the neck of the lower maxilla : it is covered externally by the parotid gland.

(2.) The *Internal Lateral Ligament* is a thin layer of considerable length, arising from the root of the styloid process and inner edge of the glenoid cavity, and attached, inferiorly, to the lower maxilla, near the orifice of the dental canal, where it is broad and membranous. It adheres to the side of the pterygoideus internus.

(3.) The *Stylo-Maxillary Ligament* also supports this articulation, extending from the styloid process to the angle of the lower jaw.

(4.) A *Membranous Capsule* surrounds the articulation, having its cavity divided into two separate bags, by

(5.) The *Inter-articular Cartilage* :—this is of an oval shape, placed transversely within the joint, irregularly convex on its upper surface, which is opposed to the glenoid cavity, while its inferior surface is hollowed, corresponding with the condyle. It separates the two synovial capsules, which are intimately united to it. This cartilage is frequently perforated, and it is adherent to the external lateral ligament, and it also gives attachment to some of the fibres of the pterygoideus externus.

SECTION II.

OF THE ARTICULATIONS OF THE SHOULDER AND UPPER EXTREMITY.

§. 1. Articulation of the Clavicle with the Sternum, or Sterno-clavicular Articulation.

The Clavicle and Sternum are articulated by their two surfaces, which are encrusted with cartilage, and connected by ligaments.

(1.) An *Anterior Ligament*, consisting of a broad band, which passes from the clavicle to the sternum over the fore part of the articulation.

(2.) The *Posterior Ligament* is smaller, and passes betwixt the two bones posteriorly.

(3.) The *Interclavicular Ligament* extends, as a flattened band, between the contiguous extremities of the two clavicles along the upper margin of the sternum, to which it adheres more or less firmly.

(4.) The *Costo-clavicular* or *Rhomboid Ligament*, is a short, broad, and strong ligament, passing from the cartilage of the first rib to a rough surface near the sternal extremity of the clavicle. It is in great part covered by the subclavius muscle.

(5.) The *Interarticular Cartilage* is observed on opening the articulation.

(6.) The *Synovial Capsule* consists of two parts which correspond to the two surfaces of the interarticular cartilage, and to the extremities of the sternum and clavicle.

This articulation is readily-examined, being nearly subcutaneous.

§. 2. Articulation of the Clavicle with the Scapula, or Scapulo-clavicular Articulation.

The plane articular surfaces of the Clavicle and of

4

the ACROMION are opposed to each other, and are connected by,

(1.) A *Synovial Capsule*, within which there is sometimes found a small interarticular cartilage.

(2.) A *Superior Ligament* passing between the two bones, covering the upper part of the articulation, and itself covered by the interlaced aponeuroses of the deltoid and trapezius.

(3.) An *Inferior Ligament* passing over the inferior surface of the articulation, continued anteriorly with the preceding ligament, but separated from it behind : below, in contact with the supra-spinatus muscle.

(4.) The *Coraco-clavicular Ligament*, which does not properly belong to the articulation, but serves to tie the clavicle to the coracoid process,—consisting of two portions, separated by an angular cellular interval. *(a.)* The *Conoid Ligament*, (the posterior and internal portion) running from the root of the coracoid process to the tubercle on the inferior surface of the clavicle, covered anteriorly by the subclavius muscle; its shape is that of an inverted cone, of which the apex is fixed to the coracoid process, the base to the clavicle. *(b.)* The *Trapezoid Ligament*, (the anterior and external portion,) of a quadrilateral form, passing from the upper part of the coracoid process to an oblique ridge on the under surface of the clavicle, and extending outwards to near the scapular extremity of the bone.

The SCAPULA has two proper ligaments.

(1.) The *Coracoid*, or *Posterior Ligament*, a thin fasciculus, extending from the base of the coracoid process across the semilunar notch of the scapula, and converting it into a foramen.

(2.) The *Acromio-coracoid Ligament*, or *Anterior Ligament of the Scapula*, is a strong, triangular, flattened ligament, stretched transversely between the coracoid process and acromion. It arises broad from the external border of the coracoid process by two fasciculi, which unite; and, thus becoming narrower but thicker, it is fixed, by its apex, to the extremity of the acromion, completing the arch formed by these two processes over the head of the humerus. This ligament has its upper surface covered by the clavicle, and by the deltoid, and its anterior border is continued into the layer of dense cellular tissue subjacent to that muscle. Its inferior surface covers the supra-spinatus.

These ligaments of the clavicle and scapula are exposed by removing the muscles, chiefly the deltoid and pectoralis major, and raising the clavicle.

§. 3. Of the Shoulder-joint, or Articulation of the Humerus with the Scapula.

This joint is also denuded, in front, by the removal of the deltoid and pectoral muscles, but, to expose it fully, the clavicle should be wholly detached, and the muscles, coming from the scapula to the head of the humerus, cleared away: their tendons are found to adhere closely to the capsular ligament of the joint.

The HEAD of the HUMERUS is received partially into the GLENOID CAVITY of the SCAPULA.

(1.) The two bones are maintained in relation by the *Capsular Ligament*, a fibrous bag, which is attached above to the circumference of the glenoid cavity, and below to the neck of the os humeri; it is lined by synovial membrane, is very loose, and, at its under part, is perforated to give passage to the biceps tendon. It is covered, at its upper part, by a dense layer of ligamentous fibres, sometimes described as (2.) A distinct *Coracohumeral Ligament*, passing from the external border of the coracoid process to the great tuberosity of the humerus, and there intermixing with the tendon of the infra-spinatus. The capsule is further strengthened, above, by the supra-spinatus,—on its outer side, by the tendons of the infra-spinatus and teres minor, and, on its inside, it is intimately united with the tendon of the subscapularis.

(3.) Cut open the joint: observe that the *synovial membrane* lines the whole surface, and is also extended into the bicipital groove, and is then reflected upwards round the tendon of the biceps, so that the tendon, although within the capsular ligament, is external to the synovial membrane.

(4.) The *Glenoid Ligament* is a projecting fibro-cartilaginous ring, which is attached to the circumference of the glenoid cavity, deepening it, and seems in part formed by the fibres of the tendon of the biceps, which bifurcates to encircle the cavity. The glenoid cavity is, however, very shallow, but the joint is strengthened by the projecting processes of the scapula, and by the ligaments and muscles.

§. 4. Of the Elbow-joint, or Articulation of the Humerus with the Radius and Ulna.

The inferior extremity of the Humerus is articulated with the superior extremities of the Ulna and Radius, the bones presenting eminences and cavities, which are reciprocally adapted to one another, forming a ginglymus or hinge-joint. The radius is also articulated with the ulna, the inner side of the circumference of its head being received into the lesser sigmoid cavity of the ulna.

The Ligaments are,

(1.) The *Internal Lateral Ligament*, fixed above to the inner condyle of the os humeri, descending over the synovial capsule, and attached, inferiorly, to the inner side of the coronoid process and also to the inside of the olecranon. It is of a triangular shape, narrow above, expanding below, and divisible into two fasciculi, which pass to the two points of insertion. The anterior fasciculus is covered by the tendinous origin of the flexor muscles, the posterior is in contact with the triceps extensor and flexor ulnaris.

(2.) The *External Lateral Ligament*, less distinct, intermixed with the common extensor tendon, and particularly with the tendon of the supinator brevis, passing from the outer condyle, over the capsule of the joint, to be lost on the coronary ligament of the radius.

(3, 4.) The *Anterior* and *Posterior Ligaments*, consisting of irregular fibrous membranes, placed one in front, the other behind the capsule of the joint, covering it partially.

(5.) The *Synovial Capsule*, extending from the articular surfaces of the humerus, and embracing the eminences and articulating cavities of the ulna and radius, and also passing into the radio-cubital part of the articulation. It lines the ligaments of the joint, and tendons adhering to it.

(6.) The *Orbicular* or *Coronary Ligament*, surrounding the neck of the radius, without adhering to it, and fixed to the anterior and posterior margins of the lesser sigmoid cavity of the ulna. It thus forms a ring in which the neck of the radius turns : it is covered by the external lateral ligament and several muscles.

The dissection of the elbow-joint only requires the cautious removal of the muscles covering it.

§. 5. Articulations between the Radius and Ulna, or Radio-Cubital Articulations.

The two bones of the fore-arm are articulated by their

extremities; and they are united, at their middle part, by intermediate ligaments.

1. The SUPERIOR RADIO-CUBITAL ARTICULATION is formed by the head of the radius and lesser sigmoid cavity of the ulna, and is maintained by the *Orbicular Ligament.* It has already been described as part of the elbow-joint.

2. The two bones do not touch in their middle portions, but are connected by,

(1.) The *Interosseous Ligament,* a thin fibrous membrane, stretched across between their opposite edges, covered anteriorly by the flexor profundus, flexor pollicis and pronator quadratus, posteriorly by the supinator brevis and proper extensors of the thumb and index. It begins below the bicipital tubercle of the radius, thus leaving at its upper part an interval for the passage of the posterior interosseous vessels; and, at its lower part, it has another oval opening for the anterior interosseous vessels and nerve.

(2.) The *Round* or *Oblique Ligament,* (or *Chorda Transversalis Cubiti,*) is a narrow band, passing from the base of the coronoid process of the ulna to the lower part of the tubercle of the radius, distinguished by the direction of its fibres, which are opposed to those of the interosseous ligament.

3. The INFERIOR RADIO-CUBITAL ARTICULATION is formed by the head of the ulna being received into a concave lateral surface of the inferior extremity of the radius. The bones are connected by

(1.) Some indistinct *anterior* and *posterior* ligamentous fibres.

(2.) An *Inter-articular Cartilage,* placed transversely between the radius and ulna, thin, narrow and triangular, adhering by its base to the radius, and attached

s 5

loosely by its apex to the styloid process of the ulna; its smooth upper surface is in contact with the inferior extremity of the ulna, while its lower surface is opposed to the cuneiform bone in the joint of the wrist.

(3.) A loose *Synovial Capsule*, (Membrana Capsularis: Sacciformis,) which passes from the radius to the ulna, and also covers the upper surface of the interarticular cartilage.

§. 6. Of the Wrist-joint, or Radio-Carpal Articulation.

This Joint is formed between the RADIUS and ULNA and the three first bones of the upper row of the CARPUS.. The scaphoid and lunar bones are received into the articular cavity of the radius, while the cuneiform bone is opposed to the under surface of the Interarticular Cartilage, which is interposed between that bone and the ulna.

The Ligaments are,

(1.) The *External Lateral Ligament*, passing from the apex of the styloid process of the radius, to the outer part of the scaphoid bone and to the trapezium.

(2.) The *Internal Lateral Ligament* from the styloid process of the ulna to the cuneiform bone, sending some fibres to the anterior annular ligament of the wrist, and to the os pisiforme.

(3, 4.) *Anterior* and *Posterior Ligaments*, broad and flat bands, which proceed from the upper border of the articulation to the first row of carpal bones, inclining obliquely inwards.

(5.) A *Synovial Membrane*, which embraces the articulation, lining the ligaments, and projecting between them.

The Wrist-Joint, and inferior radio-cubital articulation

10

should be examined together, the interarticular cartilage being common to both.

§. 7. ARTICULATIONS OF THE HAND.

The Bones of the CARPUS are placed in two rows; the first consisting of the scaphoid, lunar, cuneiform and pisiform bones : the second of the trapezium, trapezoides, magnum and unciform bones. The bones of each row are articulated mutually, and the two rows with each other.

1. The three first CARPAL BONES of the first or CUBITAL Row are articulated with each other by lateral plane surfaces, and are bound together by several ligamentous bands. The dense short fibres found in the intervals between the bones have been named *Interosseous* ligaments, while the *Dorsal* and *Palmar* ligaments extend transversely or obliquely across from one bone to the other, both in front and behind.

2. All the bones of the SECOND or DIGITAL Row are, in like manner, articulated, and united by *dorsal* and *palmar* ligaments, and by *interosseous* ligaments, of which there are two, one between the trapezoides and os magnum, and the other between the latter bone and the unciforme.

3. The Two Rows of CARPAL BONES are articulated with each other : the scaphoides with the trapezium and trapezoides, by contiguous flat surfaces ;—the cuneiform bone in the same manner with the unciforme ;—while the head of the os magnum is received into a socket, formed by the lunar and scaphoid bones. They are supported by *Anterior* and *Posterior Ligaments*, consisting of oblique fibrous fasciculi, and by two short *Lateral Ligaments ;* one external, extending from the scaphoides to the trapezium ; the other internal, passing

from the os cuneiforme to the unciforme. A *Synovial Membrane* also connects the bones of the first and second row, extends itself by prolongations between the adjacent bony surfaces, and is continued to the articulations with the Metacarpus.

4. The Os PISIFORME is peculiarly articulated with the os cuneiforme, projecting from the cubital row, and it has a separate synovial capsule, with two distinct ligaments. *(a,)* The *External* passing to the hook-like process of the unciform bone. *(b,)* The *Internal* to the upper part of the metacarpal bone of the little finger.

The bones of the carpus are also firmly connected by the *annular ligaments* of the wrist, before described, which give passage to the extensor and flexor tendons, particularly by the strong anterior or internal ligament.

5. The CARPAL BONES, in their articulations with the four last metacarpal bones are supported by *posterior* or *dorsal,* and *anterior* or *palmar ligaments,* and have synovial capsules. The articulation of the trapezium with the metacarpal bone of the thumb has a distinct *capsular ligament,* lined with synovial membrane, and strengthened by the muscles of the thumb.

6. The four last METACARPAL BONES are connected together, (1.) At their *carpal ends,* by contiguous flat surfaces, and ligamentous dorsal, and palmar bands, passing transversely. (2.) Their *digital extremities* are not in contact, but are united by a strong *Palmar Ligament,* which extends transversely, and allows of some motion.

7. The HEADS OF THE METACARPAL BONES are articulated with the concave extremities or bases of the first phalanges of the thumb and fingers, and the phalanges are articulated with each other. Each Joint is provided with (1.) An *Anterior Ligament,* which embraces the

front of the articulation, and is intermixed with the fibrous sheath of the flexor tendons. (2.) *Lateral Liga-ments,* distinct, on each side of the joint. (3.) A loose *Synovial Capsule.* The joints are strengthened behind by the extensor tendon and general tendinous expansion.

. In the first joint of the thumb, the two small *Sesamoid Bones* are found enveloped in the substance of the ante-rior ligament.

SECTION III.

OF THE ARTICULATIONS OF THE LOWER EXTREMITY.

§. 1. OF THE HIP-JOINT.

IN this joint, the HEAD of the FEMUR is received into the COTYLOID CAVITY of the Os ILIUM, forming a true Enarthrosis, or ball and socket-joint.

The hip-joint is deep-seated in the upper part of the thigh ; anteriorly, its situation is about an inch below the middle of Poupart's ligament, immediately behind the Inguinal artery : posteriorly, it corresponds with the hollow between the tuber ischii and trochanter major, situated, however, somewhat higher up, towards the dorsum of the ilium. The Joint is covered in front, by the rectus, psoas, and iliacus muscles ; on the inside, by the obturator externus and pectineus ; behind, it is in contact with the quadratus, gemelli, pyriformis, and ob-turator tendons ; and, above, it is situated beneath the gluteus minimus. These muscles must be dissected off to expose the ligaments.

. 1. The *Capsular Ligament,* consisting, (like the cap-sule of the shoulder,) of a fibrous membrane, with a synovial lining, strong and thick, embracing the whole joint, as with a circular bag ; it springs, above, from the circumference of the acetabulum or cotyloid cavity, a little beyond its margin ; below, it is inserted, anteriorly into the oblique line, which descends from the greater to the less trochanter; posteriorly, into the middle of the neck of the femur, considerably above the posterior inter-trochanteric line. This capsule is strongest at the upper and fore part, where it receives a fibrous band from the anterior inferior spinous process of the ilium : it also receives some fibres from the obturator foramen, and it is supported by the muscles which pass over it.

2. *Ligamentum Teres*, or the *Inter-articular ligament*, is a triangular chord, found within the articulation, extending from the bottom of the acetabulum to be fixed into the fossa in the head of the femur; its attachment to the acetabulum is bifurcated, consisting of two fasciculi, which arise from each extremity of the notch in the acetabulum.

3. The *Cotyloid Ligament*, or *ligamentum labri cartilagineum*, is a broad fibrous ring, which encircles and deepens the acetabulum, filling up the breaches in its edge ;—the great notch has also two strong *transverse ligaments* stretched across it.

4. *Synovial Membrane* lines the whole of the joint internally. This membrane may be traced, passing outwards from the cotyloid cavity, over both surfaces of the cartilaginous brim, then turning downwards to line the capsular ligament : reaching the femur, it is reflected upwards over the periosteum of the neck, next on the cartilaginous surface of the head, and thence it is continued round the ligamentum teres, (which is, in fact, external to it,) to the bottom of the acetabulum. It also covers the fatty substance, or *Synovial Glands*, found in the *fossa* of the acetabulum : this fatty mass receives many vessels, which enter by the great notch of the cavity.

§. 2. OF THE KNEE-JOINT, OR ARTICULATION OF THE FEMUR WITH THE TIBIA AND PATELLA.

The bony surfaces engaged in this Joint are the CONDYLES of the FEMUR, the articular surfaces of the HEAD of the TIBIA, and the posterior surface of the PATELLA. It is a ginglymus or hinge-joint, but the most complicated of all the articulations from its inter-articular ligaments and cartilages.

The Ligaments are,

(1.) The *Ligament of the Patella*, a strong ligament, of a shining tendinous appearance, and, in reality, a continuation of the extensor tendons, extending from the apex of the patella, and depression behind it, to the anterior tubercle of the tibia. Its edges are united with the tendinous expansion of the vasti; anteriorly, it is covered by the skin and fascia lata; its posterior surface is separated from the synovial capsule by fatty substance, and from the front of the tibia by a small bursa.

(2, 3.) The *External Lateral Ligaments* are two in number: the *first*, or *Long External Ligament* is a strong, round, fibrous chord, passing from the external condyle of the femur to the head of the fibula; it adheres to the external semilunar cartilage, and is covered, in great part of its extent, by the tendon of the biceps flexor. The *second* or *Short Ligament* is smaller, and descends behind the former from the back part of the condyle to the fibula.

(4.) The *Internal Lateral Ligament*, flat and expanded, broader below than above, descends from the inner condyle to the internal surface of the head of the tibia; it is attached to the internal semilunar cartilage, and it is covered below by the web-like expansion of the inner ham-string tendons.

(5.) The *Posterior Ligament* is an irregular fibrous band, placed deep behind the joint, and extending obliquely from the inner tuberosity of the tibia to the outer condyle: it is covered by the aponeurotic expansion sent off from the semimembranosus muscle, while its anterior surface is separated from the posterior crucial ligament by fatty tissue and some vessels.

To shew the *internal* ligaments of the Joint, cut freely through the capsule above the patella, and turn that bone downwards. You observe that the patella is surrounded at its apex by a large quantity of fatty substance, on each side of which is a longitudinal *Membranous fold*, chiefly formed of synovial membrane: these are termed the *Ligamenta Alaria ;* and proceeding from the lower middle part of this fatty mass to the deep fossa between the condyles, you remark another membranous duplicature, or fold of the synovial lining, which has been named the *Adipose* or *Mucous Ligament*. Divide this membranous fold, and more deeply-seated you discover the *Crucial Ligaments,* proceeding from the spine of the tibia to the back part of the depression between the condyles; and, on each side, the *Semilunar Cartilages*, with their sharp edges, partially covering the articular surfaces of the head of the tibia.

(6, 7.) The *Crucial Ligaments* are two, strong, fibrous chords, situated at the back part of the articulation, and decussating each other. The *Anterior*, arising from the inner part of the External condyle, descends, being slightly twisted, to be fixed into the anterior depression of the spine of the tibia. The *Posterior*, arising from the outer side of the Internal condyle, descends obliquely backwards, crossing behind the anterior ligament, and, spreading out, is fixed by two fasciculi, into the rough depression at the back part of the spine of the tibia, and into the external semilunar cartilage.

(8, 9.) The two *Semilunar Cartilages* are placed on the outer part of the articulating surfaces of the tibia, being interposed between them and the condyles of the femur:—flattened, and crescent-shaped, their outer convex border is thick, tapering to the inner concave edge, which is thin and sharp; both cartilages adhere by their great circumference to the lateral ligaments of the articulation, while their inner falciform edges lie unattached in the cavity of the joint, but are fixed at their extremities by fibrous bands to the head of the tibia and to the crucial ligaments, and there is generally a small *transverse Ligament* connecting the anterior extremities of the two cartilages. The internal cartilage is of a semicircular form; the external forms nearly a complete circle.

(10.) The *Synovial Capsule* extends from the shaft and condyles of the femur, above, to be attached, below, around the head of the tibia, lining the whole internal surface of the joint. It is also reflected over the articular surfaces and semilunar cartilages, and around the crucial ligaments, which are thus external to the synovial sac. Anteriorly, this capsule ascends for some way in front of the femur, lining the tendon of the extensor muscles and the aponeurosis, which covers the side of the articulation: at the back part of the joint, it surrounds the tendon of the popliteus, and is reflected before the tendons of the gastrocnemius.

§. 3. Of the Articulations between the Tibia and Fibula, or Peroneo-Tibial Articulations.

The two bones are united at their extremities by distinct articulations, and, in the middle, by the intervention of the interosseous ligament.

1. The Superior Articulation is formed by the contact of two flattened cartilaginous surfaces, one belonging to the tibia, the other to the fibula, which are connected by (1.) An *Anterior Ligament*, covered and strengthened by the biceps tendon. (2.) A *Posterior*

Ligament, covered by the popliteus. (3.) A *Synovial Membrane*.

2. In the middle of the leg, the tibia and fibula are united by the *Interosseous Ligament*, which is stretched between the opposite edges of the two bones, leaving, superiorly, a considerable interval for the passage of the anterior tibial vessels, and, below, another aperture for a branch of the peroneal artery. Its anterior surface is covered by the tibialis anticus and extensors; its posterior surface gives origin to the tibialis posticus and flexor longus pollicis.

3. The INFERIOR ARTICULATION is formed by the adaptation of the lateral articular surfaces of the tibia and fibula, over which a reflection of the synovial membrane is continued from the ankle-joint. The Ligaments are,

(1.) An *Anterior Ligament*, passing from the inferior extremity of the fibula obliquely upwards, to be fixed to the adjacent part of the tibia. It is placed in front of the two bones, on the side of the ankle-joint, the cavity of which it deepens. It is covered by the peroneus tertius muscle.

(2.) The *Posterior Ligament* is stretched between the two bones behind, and is composed of two fasciculi; the lower band is strong and distinct, crossing transversely, and forming part of the articular cavity to receive the astragalus.

(3.) Some dense, short fibres, placed between the two bones, at the upper part of the articulation, have been termed the *Interosseous Ligament*.

This Inferior Articulation should be examined in connexion with the Ankle-joint.

§. 4. OF THE ANKLE-JOINT, OR ARTICULATION OF THE TIBIA AND FIBULA WITH THE TARSUS.

The inferior extremities of the TIBIA and FIBULA form

a cavity, into which the pulley-like articular surface of the Astragalus is received. This cavity is deepened by the two malleoli, and by the ligaments of the peroneo-tibial articulation just described. The bones are maintained in situation by the following ligaments :

(1.) On the inner side of the joint, thére is one broad, quadrilateral band, the *Internal Lateral Ligament*, by some described as a *Deltoid Ligament*, descending from the inferior border of the malleolus internus to the inner part of the astragalus and of the os calcis, sending some fibres to the sheath of the long flexor of the toes. The tendon of the tibialis posticus passes over its lower part.

(2.) The *External Lateral Ligament* * is narrow, strong, and of a rounded form, lying under the tendon of the peroneus longus ; it descends perpendicularly from the lower part of the external malleolus to the outer side of the os calcis.

(3.) The *Anterior Peroneo-tarsal Ligament* † is placed in front of that last described, but is smaller, and sometimes divided into two bands : it passes from the fore part of the outer malleolus, nearly transversely forwards, to the upper and outer part of the astragalus.

(4.) The *Posterior Peroneo-tarsal Ligament* ‡ is deep-seated, and to be found at the back part of the joint, arising from the hollow behind the outer malleolus, and crossing obliquely downwards and inwards to the back part of the astragalus.

(5.) The *Tibio-tarsal* ligament consists of some irregular fibres, which pass over the fore part of the joint, from the lower extremity of the tibia to the fore part of the astragalus.

(6.) The *Synovial Membrane* lines the ligaments and

---

* *Ligamentum fibulæ medium perpendiculare* of Weitbrecht.

† *Ligamentum fibulæ anterius.* Weitbrecht.

‡ *Ligamentum fibulæ posterius.* Weitbrecht.

surfaces of the joint, and also extends to the articulation between the tibia and fibula; it is loose and contains much synovia.

§. 5. ARTICULATIONS OF THE FOOT.

The Bones of the TARSUS may be divided into two
sets or ranges; the *Posterior*, formed by the astragalus
and os calcis; the *Anterior*, by the navicular, cuboid,
and three cuneiform bones. The Bones of each division are articulated mutually, and the two ranges with
each other.

*a. Articulation of the Bones of the Posterior range;
viz.*

OF THE OS CALCIS WITH THE ASTRAGALUS. The
inferior surface of the astragalus is articulated, in two
places, with corresponding surfaces in the upper part
of the os calcis; there is a loose *Synovial Membrane*,
which is separated behind from the tendo Achillis by much
adipose substance;—and there are three ligaments:

(1.) The *Interosseous Ligament*, composed of numerous strong fibres, which pass from a depression between
the two articular surfaces of the os calcis, to a similar
groove in the astragalus.

(2.) The *Posterior Ligament* passes from the back
part of the astragalus to the neighbouring part of the
os calcis, it consists of short parallel fibres, intermixed
with the sheath of the long flexor tendon of the great
toe, which covers it.

(3.) The *External Ligament* is a rounded fasciculus,
which descends from the lower and outer part of the
astragalus to the external surface of the os calcis; it is
found on the outer side of the foot, in front of and parallel to the external lateral ligament of the ankle.

The connexion of the two bones is further strengthened by the lateral ligaments of the ankle-joint.

β. Articulations between the Posterior and Anterior Ranges of the Tarsal Bones.

A transverse articular line is observed, at the distance of an inch before the ankle-joint, dividing the tarsus into two parts; it is formed by the union of the astragalus with the navicular bone, and of the os calcis with the os cuboides. The connecting ligaments and articulations between the two ranges of bones are the following.

1. OF THE OS CALCIS WITH THE OS NAVICULARE. The two bones are not in contact, but are connected by two strong ligaments.

(1.) The *Inferior Ligament* (ligamentum calcaneo-scaphoideum inferius) is a flat, thick, fibro-cartilaginous, ligament, passing from the under and fore part of the os calcis to the under surface of the os naviculare : it is (in the erect position) immediately above the strong tendon of the tibialis posticus, and it forms with the two bones a cavity to receive the head of the astragalus. This ligament is remarkable from its dense cartilaginous nature, and is frequently called *Trochlea Cartilaginea.*

(2.) The *External calcaneo-scaphoid Ligament* consists of some very short strong ligamentous fibres, passing from the inner and front part of the os calcis to the adjacent outer portion of the navicular bone, completing the external part of the cavity for the astragalus.

2. OF THE ASTRAGALUS WITH THE OS NAVICULARE. The head of the astragalus is received in the cavity, formed by the posterior surface of the os naviculare, the os calcis, and the two ligaments just described : it forms a ball and socket-joint; there is a lining synovial membrane, and one broad ligament on the dorsum of the foot.

(1.) *Ligamentum astragalo-scaphoideum,* passing from the neck of the astragalus to the upper part of the os naviculare. It is covered by the extensor tendons.

3. Of the Os Calcis with the Os Cuboides. The anterior surface of the os calcis is articulated with the posterior surface of the os cuboides, the two bones are opposed to each other by nearly flat surfaces; there is a synovial membrane and two ligaments.

(1.) The *Superior Calcaneo-cuboid Ligament*, on the dorsum of the foot, covered by the tendon of the peroneus tertius, is a broad and thin ligament, usually divided into two or three fasciculi, passing from the anterior part of the os calcis to the corresponding portion of the os cuboides.

(2.) The *Inferior*, or *Ligamentum Longum Plantæ*, is the longest, and also the strongest of the ligaments of the foot, consisting of shining parallel fibres, and occupying the middle of the sole of the foot. It springs from the under flat surface of the os calcis, and, passing directly forwards, is fixed, anteriorly, in part to the inferior surface of the os cuboides, and by its remaining fibres, which are much longer, to the tarsal extremities of the third and fourth metatarsal bones. There is another deeper band of fibres, separated in general by fatty tissue from the ligamentum longum, which is sometimes described as a distinct ligament, passing from the os calcis to the os cuboides. The tendon of the Peroneus longus is seen crossing over the anterior extremity of this ligament.

γ. *Articulations between the Bones of the Anterior Range.*

1. Of the Os Naviculare and Os Cuboides. The external border of the os naviculare is in apposition with a small extent of the inner surface of the os cuboides. The two bones are connected by (1.) A *Dorsal Ligament*, passing somewhat obliquely from the navicular bone to the upper surface of the cuboid bone. (2.) A *Plantar Ligament*, passing transversely. (3.) Some

short *interosseus fibres;* and (4.) When the two bones
are in contact, which is not always the case, there is a
small synovial capsule and articulating surfaces.

2. OF THE THIRD CUNEIFORM BONE WITH THE OS
CUBOIDES. The external smooth surface of the former
bone is articulated with the latter, and there is here (1.)
A *Synovial Capsule.* (2.) A *Dorsal Ligament.* (3.)
A *Plantar Ligament,* stronger, and passing transversely.

3. OF THE OS NAVICULARE WITH THE THREE CUNEI-
FORM BONES. The three cuneiform bones are articulated
with distinct surfaces of the os naviculare, and are supported by ligamentous bands, which are distinguished
into (1.) *Dorsal,* and (2.) *Plantar* Ligaments: and
which may be divided in general into three portions,
above and below. There is (3.) A *Synovial Membrane,*
which is also common to the articulations of the cuneiform bones with one another.

4. OF THE CUNEIFORM BONES WITH EACH OTHER.
They have connecting *dorsal* and *plantar* ligaments,
passing transversely, and a continuation of the synovial
membrane, and also some *interosseous* fibres.

ARTICULATION OF THE TARSUS WITH THE METATAR-
SAL BONES. The Metatarsal bones are articulated with
the three cuneiform bones and with the os cuboides, but
the line of union is very irregular, chiefly from the un-
equal projection of the cuneiform bones.

The *first* metatarsal bone articulates with the first
cuneiform bone; the *second* is applied to the short second cuneiform bone, but it is also wedged in between
the first and third cuneiform bones, and has corresponding lateral articular surfaces; the *third* metatarsal
bone articulates with the third cuneiform; the *fourth*
and *fifth* with the os cuboides, the base of the last metatarsal bone projecting for half an inch beyond the external border of the os cuboides.

These articulations are provided with synovial membranes, and with *Dorsal* and *Plantar* Ligaments. The *Dorsal* consist of fasciculi, passing from the upper surface of the tarsal bones to the corresponding bases of the metatarsal bones; the *Plantar* ligaments have a like arrangement, and are further strengthened by the processes of the tendon of the tibialis posticus, and by the sheath of the peroneus longus. Strong ligamentous fasciculi also pass, in the sole of the foot, between some of the bones, which do not articulate with each other, as from the internal cuneiform bone to the bases of the second and third metatarsal bones.

The METATARSAL BONES, with the exception of the first, are articulated with each other at their posterior or tarsal extremities, by small cartilaginous surfaces, which are lined by prolongations of the synovial membrane, and maintained by *dorsal* and *plantar* ligaments extending transversely, and also by some short *interosseous* fibres. Their anterior or digital extremities are not in contact, but, as in the metacarpus, are joined by a *transverse ligament*, situated on their plantar surface, and which also includes the first metatarsal bone.

The Articulations of the METATARSAL BONES with the first phalanges, and of the PHALANGES with each other, are similar to those of the metacarpus and fingers.

THE END.

Printed by R. GILBERT, St. John's Square, London.

CPSIA information can be obtained
at www.ICGtesting.com
Printed in the USA
BVHW040609150119
R9594200001B/R95942PG537774BVX1B/1/P